Studies in Computational Intelligence

Volume 1211

The series "Studies in Computational Intelligence" (SCI) publishes new developments and advances in the various areas of computational intelligence—quickly and with a high quality. The intent is to cover the theory, applications, and design methods of computational intelligence, as embedded in the fields of engineering, computer science, physics and life sciences, as well as the methodologies behind them. The series contains monographs, lecture notes and edited volumes in computational intelligence spanning the areas of neural networks, connectionist systems, genetic algorithms, evolutionary computation, artificial intelligence, cellular automata, self-organizing systems, soft computing, fuzzy systems, and hybrid intelligent systems. Of particular value to both the contributors and the readership are the short publication timeframe and the world-wide distribution, which enable both wide and rapid dissemination of research output.

Indexed by SCOPUS, DBLP, WTI AG (Switzerland), zbMATH, SCImago.

All books published in the series are submitted for consideration in Web of Science.

Tudor Barbu

Digital Image Processing, Analysis and Computer Vision Using Nonlinear Partial Differential Equations

Springer

Tudor Barbu
Institute of Computer Science
Romanian Academy—Iasi Branch
Iasi, Romania

ISSN 1860-949X ISSN 1860-9503 (electronic)
Studies in Computational Intelligence
ISBN 978-3-031-89575-3 ISBN 978-3-031-89576-0 (eBook)
https://doi.org/10.1007/978-3-031-89576-0

This Springer imprint is published by the registered company Springer Nature Switzerland AG
The registered company address is: Gewerbestrasse 11, 6330 Cham, Switzerland

This book is dedicated to all those who have helped me in my research career.

Contents

Chapter 1
Introduction

Partial differential equations (PDEs), which have long been used to express various dynamical phenomena since they describe the continuous change, have become ubiquitous in mathematically oriented scientific domains, like the engineering and physics. Thus, these equations are fundamental for the scientific understanding of the diffusion, heat, electrostatics, electrodynamics, thermo-dynamics, fluid dynamics, elasticity or quantum mechanics.

In the last 40 years the partial differential equations have been applied successfully in traditionally engineering fields, like the signal and image processing and analysis, given their advantages in both theory and computation: modelling flexibility, rigorous mathematical foundation, high speed, stability and accuracy. The PDE-based models represent very effective tools for solving digital image processing and analysis tasks such as the image denoising and restoration, inpainting, compression, segmentation, registration and video motion estimation, addressing properly the challenges that still persist in these domains.

This book is concerned with the applications of the nonlinear partial differential equations in these digital image and video processing and analysis, and even computer vision fields, describing numerous PDE-based models, in variational and non-variational form, that have been introduced in these areas in the last decades, but especially focusing on the most important contributions of the author in those domains. So, the main theoretical and practical achievements of his research in the PDE-based image filtering, inpainting, compression, segmentation, content-based recognition, indexing and retrieval, and video object detection and tracking are presented here. Energy-based (variational) and nonlinear diffusion-based models of second and fourth order proposed for all those tasks will be discussed.

This work can be regarded as a continuation of the previous author's book, *Novel Diffusion-based Models for Image Restoration and Interpolation*, Springer, 2019. Like the previous one, this book addresses to students, researchers and professionals working in mathematics, computer science and electrical engineering. An elementary

T. Barbu, *Digital Image Processing, Analysis and Computer Vision Using Nonlinear Partial Differential Equations*, Studies in Computational Intelligence 1211,
https://doi.org/10.1007/978-3-031-89576-0_1

level of knowledge of the partial differential equations theory, numerical analysis and digital image processing is the required background of the readers.

While the previous book described the applications of linear and nonlinear PDE models in 2D image denoising and interpolation, this new book deals with the nonlinear PDEs only, but addresses almost all the PDE-based image processing and analysis areas, the connections between partial differential equations, computer vision and artificial intelligence, and considers other image types also.

The book is composed of seven chapters, this introduction representing the first of them, and a section of references. The second chapter surveys the nonlinear PDE-based image denoising and restoration techniques. The PDE-based filtering of the additive white Gaussian noise (AWGN) is considered in its first section. The most important total variation (TV)-based and second- and fourth-order anisotropic diffusion-based AWGN removal solutions, including our own nonlinear parabolic and hyperbolic PDE-based AWGN-based filters, are described here. While signal-independent AWGN represents the most popular type of image noise, many digital images are contaminated by signal-dependent quantum (Poisson, shot) noise, resulting from fluctuations in the number of detected photons, or mixed Poisson-Gaussian noise during the acquisition process. Nonlinear PDE-based filtering models dealing with these types of noise are presented in the next two sections. The variational PDE, hyperbolic PDE and nonlinear anisotropic diffusion models developed by us for photon-limited image restoration and mixed noise filtering constitute the main focus of these sections. Some stochastic differential equation (SDE)-based models that lead to nonlinear PDE-based Gaussian, quantum and mixed noise reduction filters are also presented in these sections.

Nonlinear PDE-based filtering models provide effective gray-level image restoration results, removing various types of noise, overcoming unintended effects like blurring and staircasing, preserving important details such as edges or corners, and outperforming the conventional 2D filters. But besides the static 2D image denoising, the nonlinear PDEs represent effective tools for multi-channel and video image restoration. The vector-valued PDE models used for the filtering of multi-layer images, such as the multispectral (MSI), hyperspectral (HSI) and multimodal images, are discussed in the fourth section. Our own nonlinear vector-valued PDE-based multi-channel image restoration models described here lead to systems of reaction–diffusion equations sharing some correlation terms, which filter efficiently various types of noise from various types of multi-layer images, while dealing properly with the channel inter-correlation issue. The nonlinear PDE-based video denoising and restoration area is treated in the last section of the chapter, the recent PDE-based video image filters being surveyed. Our contributions in this field include the 3D nonlinear anisotropic diffusion-based models for movie sequence filtering.

Next chapters treat the connection between the nonlinear PDEs and the artificial intelligence (AI), demonstrating how the PDE + AI combination represents an useful tool for solving numerous image analysis and computer vision tasks. The third chapter is dedicated to two closely related domains, which are the PDE-based inpainting and image compression. The inpainting, also called image interpolation or completion, is also strongly related to image restoration, since the inpainting domain, which is composed of the missing regions of a digital image, can be interpolated by directing the denoising process to it. Thus, the nonlinear PDE-based filtering models could be adapted for this reconstruction task by adding inpainting masks to them.

The structural nonlinear PDE-based inpainting field is approached in the first section, where the state of the art PDE-based image interpolation techniques are described first. They include total variation-based inpainting schemes and nonlinear second-order diffusion-based inpainting models that follow variational principles, as well as some important non-variational third- and fourth-order nonlinear PDE-based inpainting approaches. The nonlinear PDE-based inpainting techniques developed by us, which use variational PDE and non-variational second- and fourth-order anisotropic diffusion models derived from PDE-based filtering schemes, are also surveyed in this section.

The nonlinear PDE-based image compression domain treated in the second section can be regarded as an application area of the structure-based inpainting methods, since the PDE-based image interpolation models can be used successfully in the decompression stage. The PDE-based compression frameworks usually perform an image sparsification process, followed by a pixel encoding procedure. They apply a pixel decoding scheme, followed by a nonlinear PDE-based scattered data interpolation, for decompression. The most important PDE-based image compression techniques, including those based on linear homogeneous diffusion and edge-enhancing diffusion (EED), such as BTTC-EED with its improved versions and the rectangular subdivision with edge-enhancing diffusion (R-EED), are surveyed first. Then, our own nonlinear PDE-based compression solutions are described. Unlike other compression methods that use PDEs only for image decompression, one of our techniques uses nonlinear PDE-based models in both compression and decompression stages. So, it applies an edge-based sparsification algorithm using a PDE-based edge extraction and a nonlinear diffusion-based inpainting model combining second and fourth order anisotropic diffusion terms. Another compression framework proposed by us provides a feature keypoint-based sparsification involving feature descriptors used in computer vision and AI, and a nonlinear parabolic fourth-order PDE-based decompression scheme.

The PDE-based multi-scale analysis forms the basis of the most image analysis and computer vision tools described in the next three chapters. The nonlinear PDE-based scale-space representations proved to be more effective than other multi-scale spaces, like those based on 2D Gaussian kernels or Gaussian derivatives. The multi-scale image analysis frameworks described in the fourth chapter use nonlinear diffusion-based scale spaces combined to AI-based descriptors in order to perform high-level feature extractions.

The first section surveys the existing multi-scale texture recognition techniques, insisting mostly on our anisotropic diffusion-based texture analysis methods. The proposed supervised and unsupervised texture recognition approaches create multi-scale spaces using the numerical approximation schemes of some nonlinear second- and fourth-order PDE models and perform texture feature extractions applying combinations of rotation-invariant descriptors at multiple scales. The content-based image recognition technique described in the second section extracts the content features by applying a bank of circularly symmetric 2D filters at each scale of a fourth-order anisotropic diffusion scheme. An overview of cluster-based image content-based indexing and retrieval (CBIR) methods is provided in the third section. Then, one describes our own multi-scale CBIR framework, composed of a cluster-based indexing component performing a high-level feature extraction, which combines deep learning neural networks to a second-order nonlinear reaction–diffusion based scale-space representation, and then a graph-based clustering, and a relevance feedback-based image retrieval mechanism.

The static and video image segmentation, representing an important image analysis and computer vision field, is successfully approached using energy- and PDE-based models in the fifth chapter. The edge-based static image segmentation methods using nonlinear PDE-based edge detection schemes are surveyed in its first section. The PDE-based multi-scale analysis constitutes an effective tool used for this task. We also developed a performant nonlinear diffusion-based multi-scale boundary extraction framework that is presented here. It detects edges at each scale of the proposed second-order parabolic PDE-based scale-space by searching for zero-crossings and applying gradient magnitude thresholding and morphological operations. The edges determined at multiple scales are combined using a fine-to-coarse edge tracking scheme into the final boundary detection output.

The active contour-based image segmentation techniques are surveyed in the next section. Those using various types of parametric Active Contours are described in the first sub-section, while the Geodesic (geometric) Active Contours (GAC) are presented in the second one. Our main contributions in this area, representing GAC-based segmentation schemes using variational and nonlinear diffusion models with level-set functions, are also detailed here. They are used successfully in combination to machine learning algorithms like the Gentle AdaBoost (GAD) cascade classifiers and Gaussian Mixture Models (GMM) for solving optimally the video object detection tasks.

The temporal video segmentation field is considered in the last section of the chapter. An overview of the temporal movie sequence segmentation methods, including our own techniques, is provided first. Then, the usefulness of the nonlinear PDE-based multi-scale analysis in this domain is proved by an effective multi-scale video shot detection framework developed by us. It creates a scale-space representation applying the iterative numerical approximation algorithm of a nonlinear second-order hyperbolic PDE with boundary conditions. A high-level convolutional neural network (CNN)-based video frame feature extraction combining Inception-ResNet-V2 and DenseNet-20 deep networks is performed at multiple scales, then an automatic clustering is applied to the distances between the feature vectors of adjacent frames.

The nonlinear PDE-based video analysis is also approached in the sixth chapter, which surveys the video object detection and tracking, a well-known computer vision domain. Thus, variational detection and tracking approaches representing some influential Active Contour models are surveyed in the first section. The presented results illustrate that the PDE-based active contours can be applied successfully for both the object detection and tracking tasks.

Since the optical flow represents a very effective solution to track the video object motion, some variational optical flow-based moving object detection and tracking techniques are surveyed in the second section. One describes here several variational PDE-based optical flow estimation models, such as the Horn–Schunck method and its improved versions, and explains their application to the detection and tracking task. Also, some moving object detection and tracking schemes that combine the optical flows to the edge information or some well-known invariant feature descriptors, are also discussed in this section.

Some of our main contributions in this computer vision field, representing pedestrian and vehicle detection and tracking frameworks, are detailed in the last chapter section. They demonstrate that the diffusion-based multi-scale image analysis and the combinations between the PDE and AI models represent effective tools for this task, too.

While the detection of the object instances in the video frames is performed by combining several machine and deep learning models, such as the GMM, SVM + HOG and Aggregate Channel Features (ACF) for moving persons, and GMM, ACF, YOLO networks and Faster R-CNN for vehicle detection, their instance-matching based tracking-by-detection (TBD) involves a high-level nonlinear PDE-based multi-scale analysis of the sub-images of the detected objects. The scale-space representations are created by applying the numerical solvers of some nonlinear second- and fourth-order reaction–diffusion models on those sub-images at some properly chosen iteration moments. The high-level feature extractions performed at multiple scales use combinations of machine and deep learning models, such as the mixtures of SURF and HOG-based descriptors, Inception-V3 and ResNet-101 convolutional neural networks, and the deep networks ResNet-50 and MobileNet-v2 respectively. Effective tracking-by-detection algorithms using the computed high-level feature vectors are provided in each case.

The main conclusions of this book are drawn in the last chapter. The future research plans in the approached fields are also laid out there. The bibliographic reference section of each chapter, except those of introduction and conclusions, lists all the published scientific works used for its preparation, many of them disseminating our research.

The nonlinear PDE-based models introduced in this book for solving various image processing and analysis tasks were investigated seriously, rigorous treatments of their mathematical validity, or well-posedness, being performed for them. The most partial differential equations introduced by us are well-posed, since the existence of a unique weak, or variational solution, of them can be demonstrated, at least if some certain assumptions are satisfied.

The solutions of these PDEs are determined numerically by applying some approximation algorithms. The proposed iterative numerical approximation schemes, created using the finite-difference method, are stable and consistent to the nonlinear PDE models and converge fast to their numerical solutions. These numerical algorithms are used successfully for solving image processing tasks like the denoising and inpainting, and to create effective multi-scale space representations for various image analysis and computer vision tasks.

Many image processing and analysis experiments and method comparison illustrating the effectiveness of the described techniques are also presented in this book. Our own algorithms are compared to the state of the art approaches of their domains and their performance is assessed by applying some rigorous performance measures related to their domains. The described method comparison results show that our nonlinear PDE-based techniques outperform or perform comparably good to numerous others well-known approaches in those fields. These results also demonstrate once more that the nonlinear partial differential equations and the artificial intelligence could work together successfully to achieve significant accomplishments.

Chapter 2
Nonlinear PDE-Based Digital Image Restoration Techniques

This chapter addresses the nonlinear partial differential equation (PDE)-based image denoising and restoration domain, detailing our most important contributions in this field. It describes PDE-based filtering models for various types of noise and images. Thus, the first three sections present 2D image restoration techniques related to different noise distributions. The first one introduces several effective nonlinear diffusion-based models, in variational and non-variational form, for additive white Gaussian noise (AWGN) reduction. The nonlinear PDE-based quantum noise filtering solutions are described in the second section, while the variational and anisotropic diffusion schemes dealing successfully with the noise mixtures are discussed in the third section. The last two sections are concerned with the restoration of other types of digital images. The fourth section describes multi-channel image denoising techniques using vector-valued PDE models, while the video image filtering solutions based on 3D nonlinear anisotropic diffusion models are presented in the fifth section.

2.1 Variational and Anisotropic Diffusion Schemes for AWGN Removal

The image denoising constitutes an essential image processing task that consists on reducing the electronic noise, representing a random variation of brightness or color information, in the image signal. There exist multiple types of noise that contaminate the digital images, the most often encountered of them being the additive white Gaussian noise (AWGN).

This common additive noise represents a basic noise model used in information theory to mimic the effect of various random processes occurring in nature, which is characterized by a probability density function (PDF) equal to that of the normal

T. Barbu, *Digital Image Processing, Analysis and Computer Vision Using Nonlinear Partial Differential Equations*, Studies in Computational Intelligence 1211,
https://doi.org/10.1007/978-3-031-89576-0_2

distribution, $p(z) = \frac{1}{\sqrt{2\pi}\sigma} e^{-(z-u)^2/2\sigma^2}$, where z is the gray level, μ stands for the mean of random variable z and σ represents its standard deviation [1, 2]. AWGN represents a signal-independent noise that comes from the spontaneous thermal generation of electrons during image acquisition and transmission processes. The formation model of this type of noise is $u^* = Hu + N(0, \sigma^2)$, where the gray-level image is expressed as a function $u : \Omega \subseteq R^2 \rightarrow R$, u^* is the observed image, H: $\mathfrak{H} \rightarrow \mathrm{K}$, where K is a real Hilbert space, represents a bounded linear operator, such as a convolution, and $N(0, \sigma^2)$ is the additive noise. A smoothing procedure must restore u from the observation u^*, but identifying an optimal denoising solution is still a challenge in this image processing field, since such a restoration technique has to optimize the trade-off between image filtering, detail preservation and unintended side-effect removal. While the conventional 2D filtering approaches generate the undesirable blurring effect and do not preserve properly the boundaries and other essential features [1, 2], the nonlinear PDE-based filters that are discussed in this research work represents a much better solution to the restoration issue. Most of these filtering solutions are based on some diffusion equations.

Diffusion represents the physics process where the molecules of a material move from a high-concentration area to a low-concentration area until an equilibrium is reached. This process, which is modeled using partial differential equations, has been applied successfully in the image processing domain.

Thus, since the diffusion process conserves the transported mass, the conservation law describing the physical transport can be applied for it [3]. Its formula is $\frac{\partial u}{\partial t} = -divJ$, where the *diffusion flux* is given by Fick's *first law of diffusion* as $J = -D \cdot \nabla u$, D stands for the *diffusion tensor* and ∇u represents the *concentration gradient* [3]. These equations lead to the following diffusion-based equation:

$$\frac{\partial u}{\partial t} = div(D \cdot \nabla u) \tag{2.1}$$

It can be applied to the image denoising and restoration task if one considers $u : \Omega \subseteq R^2 \rightarrow R$ as the evolving image of an initial observation u_0 that is contaminated by AWGN. The type of the diffusion-based restoration process given by (2.1) depends on the tensor, or conductance diffusivity function, $D\ (x, y, t)$. Thus, if D does not depend on u itself, then (2.1) becomes a *linear diffusion* equation. Otherwise, if the tensor is a function of u, (2.1) represents a *nonlinear diffusion* equation. Also, if D is constant over the entire domain ($D(x, y, t) = \alpha, \forall (x, y) \in \Omega$), one obtains a *homogeneous*, or *isotropic diffusion*. If D becomes a space-dependent function, then (2.1) represents an *inhomogeneous*, or *anisotropic*, *diffusion* equation [4].

The most known linear diffusion equation is the *heat equation*, which describes the temperature variation's distribution [5]. The image filtering process provided by it is equivalent to that of a 2D Gaussian filter and has the same drawbacks. Although there are some linear PDE-based models that provide quite good restoration results [6], they are clearly outperformed by the nonlinear PDE-based filtering techniques that are described in the next two subsections.

2.1.1 Nonlinear Second-Order PDE-Based Gaussian Noise Removal Methods

The nonlinear diffusion-based second-order PDE filtering models provide an effective image restoration, since they remove successfully the AWGN and overcome unintended effects, like blurring. They perform the denoising process along, and not across, the boundaries, thus preserving the essential image details.

An influential second-order anisotropic diffusion-based filtering scheme is that introduced by Perona and Malik in 1987 [7]. It is given by the following nonlinear parabolic PDE:

$$u_t = \frac{\partial u}{\partial t} = div\left(g\left(\|\nabla u\|^2\right) \cdot \nabla u\right) \tag{2.2}$$

where the *diffusivity conductance*, or *edge-stopping*, function $g : [0, \infty) \to [0, \infty)$ is positive, monotonous decreasing and converges to 0. They considered two such functions, $g\left(s^2\right) = e^{-\frac{s^2}{k^2}}$; $g\left(s^2\right) = \frac{1}{1+\left(\frac{s}{k}\right)^2}$, where the conductance parameter $k > 0$. A finite difference-based numerical approximation scheme that solves this PDE model was also proposed by them [6]. The P-M scheme provides effective denoising results and, since it has a strong edge-preserving character, it has been also used successfully for the edge detection task [6].

This anisotropic diffusion model has been investigated mathematically a lot in the last 35 years and many nonlinear diffusion-based filtering solutions inspired by it and improving it have been developed in these years [4, 8]. Some of these derived techniques introduce other variants of edge-stopping functions.

Thus, the TV-ROF denoising scheme uses $g\left(s^2\right) = \frac{1}{|s|}$, whose regularized form is $g\left(s^2\right) = \frac{1}{\sqrt{s^2+\varepsilon^2}}$ [9]. The Charbonnier diffusion-based model is based on the diffusivity function $g\left(s^2\right) = \left(1 + \frac{s^2}{k^2}\right)^{-1/2}$ [10]. The Weickert diffusion approach is characterized by $g\left(s^2\right) = \begin{cases} 1 - e^{-\frac{C_m}{\left(s^2/k^2\right)^m}}, & if\ |s| > 0 \\ 1, & if\ s = 0 \end{cases}$ [11], with $1 = e^{-C_m}(1 + 2C_m m)$, $m \in \{2, 3, 4\}$, $C_2 = 2.3366$, $C_3 = 2.9183$ and $C_4 = 3.3148$. The *robust anisotropic diffusion* (RAD) model introduced by Black et al. applies the robust estimation theory in creating the next conductance function, called *Tukey's biweight*: $g\left(s^2\right) = \begin{cases} \left(1 - \frac{s^2}{5k^2}\right)^2, & if\ \frac{s^2}{5} \le k^2 \\ 0, & if\ \frac{s^2}{5} > k^2 \end{cases}$ [12].

Various choices of the conductance parameter k have also been proposed. It must be selected carefully for a satisfactory edge-preserving AWGN reduction, because, when the gradient magnitude exceeds its value, that boundary is sharpened. While some models, including Perona–Malik technique, use a fixed k that can be determined empirically, other approaches consider this parameter as a function of time, $k(t)$, that is reduced gradually during the smoothing process, or detect it automatically, as a function of the current state of the evolving u [13]. So, some detection solutions, like the morphology-based $k = avg(u \circ S) - avg(u \bullet S)$ or various statistics-based approaches [13], were introduced.

Other Perona–Malik inspired denoising methods represent some regularization attempts of the influential model, which is ill-posed. Such a well-posed regularization approach, proposed by Catte et al., applies a Gaussian filter to the evolving image: $u_t = div(g(\|\nabla u_\sigma\|) \cdot \nabla u)$, where $u_\sigma = K_\sigma * u$ [14]. Other regularization techniques were proposed by Kacur and Mikula [15] and Torkamani-Azar and Tait [16].

And some second-order PDE-based solutions improve the P-M model by introducing various fields in its scheme. They include Gradient Vector Flow (GVF)-based P-M [17], INGVF-based P-M [18], Gradient Vector Convolution (GVC)-based P-M [19], and CONVEF-based P-M models [20].

We have also developed some second-order anisotropic diffusion-based filtering models that improve and regularize the Perona–Malik approach, by introducing new conductance diffusivity functions and parameters, and new boundary conditions. Such a technique, which was proposed in [21], uses the parabolic PDE model

$$\begin{cases} \frac{\partial u}{\partial t} = div\left(\psi_{K(u)}\left(|\nabla u|^2\right)\nabla u\right) \\ u(0, x, y) = u_0 \\ \nabla u \cdot \nu = 0, \quad \text{on } (0,T) \times \partial\Omega \end{cases}, \quad (x, y) \in \Omega \tag{2.3}$$

whose edge-stopping function is $\psi_{K(u)} : [0, \infty) \to [0, \infty)$, with

$$\psi_{K(u)}(s^2) = \begin{cases} \alpha\sqrt{\frac{K(u)}{\beta s^2+\eta}}, & \text{if } s \in [0, M] \\ \frac{\alpha}{\sqrt{\gamma}}, & \text{if } s = 0 \end{cases} \tag{2.4}$$

where $\alpha, \beta \in [0.5,\ 0.8]$, $\eta \in [0.5, 1)$, $M > 0$, ν is the normal to $\partial\Omega$ and the conductance parameter is computed as the statistics-based function:

$$K(u) = \|u\|_F \frac{median(u)}{n(u)\varepsilon} \tag{2.5}$$

with $\varepsilon \in (0, 1]$, $\|u\|_F$ is the Frobenius norm of u, $median(u)$ returns its median and $n(u)$ is its number of pixels. This nonlinear diffusion model is valid, since it admits a unique variational (weak) solution. Its well-posedness is demonstrated rigorously in [21, 22]. The filtering approach based on it removes successfully the additive noise, overcome the blurring effect and outperforms many conventional and nonlinear PDE-based restoration methods [21]. An AWGN filtering method comparison example is provided in Fig. 2.1, where the restoration results provided by our technique, the P-M scheme and other filters on the *Peppers* image are displayed.

Other Perona–Malik inspired filtering schemes were introduced by us in [23, 24] and also surveyed in [22]. For example, the parabolic PDE-based filter described

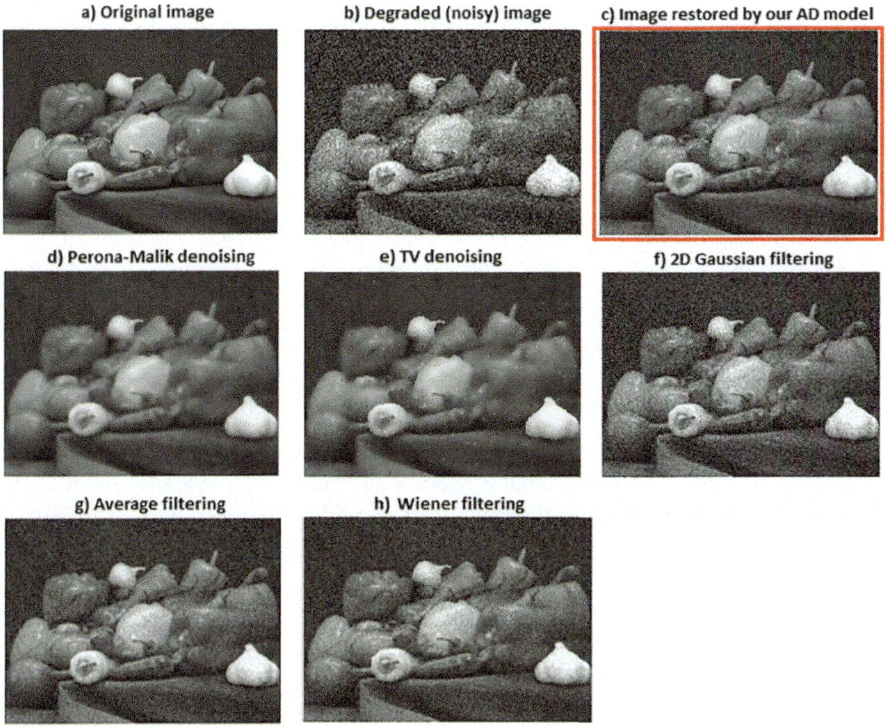

Fig. 2.1 AWGN denoising results achieved by several filters

in [24] uses the conductance function $\xi_u(s) = \frac{\alpha}{\left(\frac{s}{\eta_u}\right)^k + \eta_u \left|\ln\left(\frac{s}{\eta_u}\right)\right|}$ with the parameter $\eta_u = \beta\mu(\|\nabla u\|) + \gamma|median(u)|$, where $\beta \in (2, 3)$, $\gamma \in (1, 2)$.

An important category of nonlinear diffusion-based denoising schemes contains the *mean curvature motion* (MCM)-based PDE models. This well-known MCM-based anisotropic diffusion model is introduced by Alvarez et al. [25]:

$$\begin{cases} \frac{\partial u}{\partial t} - g(|G_\sigma * \nabla u|)\|\nabla u\| div\left(\frac{\nabla u}{|\nabla u|}\right) = 0 \\ u(0, x, y) = u_0(x, y) \end{cases}, \tag{2.6}$$

where G_σ is the 2D Gaussian filter kernel and $g : [0, \infty) \to [0, \infty)$ is a non-increasing real function satisfying $\lim_{s \to \infty} g(s) = 0$. The diffusion-based models provided by (2.6) produce an effective selective denoising and edge detection [22, 25].

The *morphological anisotropic diffusion* models include the algorithm developed by Segall and Acton [26]. Their PDE-based approach introduced a morphological diffusion coefficient that is able to restore the low-sized objects while preserving and enhancing the image edges. Another class of nonlinear PDE-based filtering schemes uses the *complex diffusion* processes. The nonlinear complex anisotropic diffusion-based filter introduced by Gilboa et al. has the form [27]:

$$u_t = \nabla \cdot (c(\mathrm{Im}(u))\nabla u) \tag{2.7}$$

where $c(\mathrm{Im}(u)) = \frac{e^{i\theta}}{1+\left(\left|\frac{\mathrm{Im}(u)}{k\theta}\right|\right)^2}$, k is a threshold value and phase angle $\theta \ll 1$.

The *edge-enhancing diffusion* (EED) based smoothing provides a much better edge preserving and enhancement while removing the image noise [4]. It is given by the following nonlinear PDE:

$$\frac{\partial u}{\partial t} = div\left(g\left(\nabla u_\sigma \nabla u_\sigma^{\mathrm{T}}\right)\nabla u\right) \tag{2.8}$$

Coherence-enhancing anisotropic diffusion represents an extension of the edge-enhancing anisotropic diffusion that is specifically designed to enhance the line-like structures by integrating the orientation information [4, 28]. It is based on a diffusion tensor containing directional information and having the following form:

$$D = [v_1, \quad v_2]\begin{bmatrix} g_{\lambda_1}(\lambda_1, \lambda_2) & 0 \\ 0 & g_{\lambda_2}(\lambda_1, \lambda_2) \end{bmatrix}\begin{bmatrix} v_1 \\ v_2 \end{bmatrix} \tag{2.9}$$

where v_1, v_2 are the eigenvector parallel and orthogonal to the image gradient respectively and λ_1, λ_2 are the corresponding eigenvalues [28], and the proposed edge stopping function is

$$\begin{cases} g_{\lambda_1}(\lambda_1, \lambda_2) = \alpha \\ g_{\lambda_2}(\lambda_1, \lambda_2) = \alpha + (1-\alpha)e^{\frac{-C_m}{\left(k/(\lambda_1-\lambda_2)^2\right)^m}} \end{cases} \tag{2.10}$$

Most second-order diffusion models follow variational principles, since it is very common to achieve nonlinear PDE-based denoising model from energy-based (variational) problems representing minimizations of some energy cost functionals [8, 29]. Such an energy functional is composed of a *regularization* term and a *fidelity* one. The generic variational denoising scheme is based on the next minimization:

$$\min_u\{E(u) = R(u) + F(u)\} \tag{2.11}$$

where $E(u)$ stands for the energy functional, the regularization term is

$$R(u) = \alpha \int_\Omega \psi(\|\nabla u\|)d\Omega \tag{2.12}$$

with regularizer function ψ and the fidelity term takes the next form:

$$F(u) = \frac{\beta}{2} \int_{\Omega} (u - u_0)^2 d\Omega \tag{2.13}$$

where u_0 is the initial noisy observation and $\alpha, \beta > 0$.

Many variational PDE-based image filtering schemes have been proposed in the last three decades. These variational models lead to nonlinear PDEs by creating the associated Euler–Lagrange equations, then applying the steepest descent method.

The *total variation* was introduced for image denoising and restoration in 1992 by Rudin et al. [9]. Their influential variational filtering model, named Total Variation (TV) Denoising or TV-ROF, consists in the minimization of the total variation norm. TV Denoising is very effective at simultaneously conserving the image edges while removing the AWGN in flat regions, and has the form:

$$u_{rest} = \arg \min_{u \in L^2(\Omega)} \int_{\Omega} \left(\|\nabla u\| + \frac{1}{2\lambda} (u_0 - u)^2 \right) d\Omega \tag{2.14}$$

with $\lambda > 0$ and u_{rest} is the restoration. The next second-order diffusion equation is obtained applying the Euler–Lagrange equation and steepest descent method:

$$\frac{\partial u}{\partial t} = div\left(\frac{\nabla u}{|\nabla u|} \right) + \frac{1}{\lambda}(u - u_0) \tag{2.15}$$

See a TV-ROF image denoising example in Fig. 2.1e. Numerous variational restoration approaches that improve this total variation regularization model, by addressing drawbacks like its vulnerability to the staircase effect, have been developed. The ROF Total Variation Denoising with Split Bregman employs an iterative Split Bregman algorithm to solve the TV minimization [30]. The TV-L_1 denoising model improves TV-ROF by using L_1-norm as a measure of fidelity between the observation and restored image [31]. It is based on the minimization:

$$\min_{u \in BV(R^N)} \int_{R^N} |\nabla u| + \lambda \int_{R^N} |u - u_0| dxdy \tag{2.16}$$

TV-L_1 is contrast invariant and able to separate the image features according to their scales [31].

Another improved variant of the classic TV model is the Adaptive TV Denoising (ATV) based on difference curvature [32]. Its regularization term operates like a TV norm at object boundaries in order to preserve the edges, while approximating to the L_2-norm in the flat and ramp image regions in order to overcome the staircasing. ATV is based on the minimization problem:

$$\min E_{ATV} = \int_{\Omega} \left(|\nabla u|^{p(D)} + \frac{1}{2} \lambda(D)(u - u_0)^2 \right) dxdy \tag{2.17}$$

where the functions $p(D) = 2 - \sqrt{\overline{D}}$, $\lambda(D) = k \cdot \sqrt{\overline{D}}$, D stands for the difference curvature of the image, $\overline{D}$ is its normalization and $k > 0$ [32].

Total Variation Denoising method based on median filter and phase consistency (MPC-TV) improves TV Denoising by modifying it using a diffusion rate adjuster using the phase congruency and a fusion of the median filter and phase consistency boundary [33]. This TV-based filtering approach removes successfully the AWGN and deals properly with speckle noise also.

The Generalized Total Variation (GTV) regularization-based image noise removal, constitutes another improved TV Denoising scheme [34]. GTV Denoising uses a generalized pth power total variation and has an improved performance.

We also derived some nonlinear second-order parabolic PDE-based filters from energy-based models. Such a variational denoising technique proposed by us minimizes the energy functional [35]:

$$E(u) = \int_{\Omega} \left(\frac{\lambda}{2} \psi_u(\|\nabla u\|) + \frac{\rho}{2} (u - u_0)^2 \right) d\Omega, \tag{2.18}$$

where $\lambda, \rho \in (0, 1)$ and its regularizer is

$$\psi_u(s) = \int_0^s \tau\zeta \left(\frac{\gamma(u)}{\beta \ln(s + \gamma(u))^3 + \delta} \right)^{1/2} d\tau, \tag{2.19}$$

whose conductance parameter $\gamma(u) = \alpha\mu(\|\nabla u\|) + \eta t(u)$, the parameters $\alpha, \beta, \zeta, \eta, \delta \in (0, 3]$ and $t(u)$ returns the current time of u in evolving sequence. A nonlinear anisotropic diffusion-based model is obtained by applying the Euler–Lagrange equation and gradient descent method [36]. A finite difference method-base discretization scheme was created for its numerical solving [35]. It outperforms many nonlinear PDE-based filters and classic denoising schemes, achieving higher scores of the performance metrics. Other variational denoising models leading to nonlinear second-order diffusion equations were introduced in [37].

While the most existing nonlinear PDE-based restoration models are parabolic can be derived from some variational schemes, we introduced effective denoising approached based on non-variational PDEs that are also not parabolic. Thus, several hyperbolic PDE-based filters that contain also nonlinear diffusion components were developed by us, disseminated in our past papers and also surveyed in [22]. They reduce considerably the AWGN and sharpens the image details, like edges and corners, given the second order derivatives used by them. So, one of these models is based on the following hyperbolic PDE with boundary conditions [38]:

$$\begin{cases} \alpha \frac{\partial^2 u}{\partial t^2} + \beta^2 \frac{\partial u}{\partial t} - div\left(\psi_{K(u)}(\|\nabla u\|)\nabla u\right) + \lambda(u - u_0) = 0 \\ u(0, x, y) = u_0(x, y) \\ \dfrac{\partial u}{\partial t}(0, x, y) = u_1(x, y) \\ u(t, x, y) = 0, \quad \forall t \geq 0, \quad (x, y) \in \partial\Omega \end{cases}, \quad (x, y) \in \Omega \tag{2.20}$$

with $\alpha, \beta \in (0, 1]$ and $\lambda \in (0, 0.4]$ and the edge-stopping function modeled as:

$$\psi_{K(u)}(s) = \begin{cases} \xi\sqrt{\frac{K(u)}{\gamma s^2 + \eta}}, & \text{if } s > 0 \\ 1, & \text{if } s = 0 \end{cases} \tag{2.21}$$

where $\xi, \gamma \in (0, 1]$, $\eta \in (0, 6)$ and conductance $K(u) = \zeta\,|median(\|\nabla u\|) - \varepsilon| + \nu ord(u)$, where $\zeta, \varepsilon \in (0.2{,}3]$, $\nu \in (0, 0.6)$, $ord(u)$ returns the order of current state of u. It provides very good edge-enhancing denoising results, avoiding the side-effects [38].

2.1.2 Nonlinear Fourth-Order PDE-Based AWGN Filtering Approaches

While the nonlinear second-order PDE-based filters remove successfully the AWGN, avoid the blurring and preserve well the image edges, they use to approximate the observations as step images looking *blocky* and usually generate the staircase effects. For this reason, the nonlinear fourth-order partial differential equations were introduced as an effective denoising and restoration tool that solve the drawbacks of the second-order PDE models. Unlike those nonlinear second-order diffusion schemes, the fourth-order PDE-based models generate piecewise planar images that look more natural and overcome the unintended staircasing [22].

An influential nonlinear fourth-order PDE-based restoration approach is the isotropic diffusion-based scheme proposed by You and Kaveh in 2000 [39]. Their L_2-curvature gradient flow method is achieved from a variational problem based on the next energy functional minimization:

$$E(u) = \int_{\Omega} f\left(\left|\nabla^2 u\right|\right) dxdy \tag{2.22}$$

with f an increasing function. It leads to the nonlinear 4th-order PDE-based model:

$$\begin{cases} \frac{\partial u}{\partial t} = -\nabla^2\left[g\left(\nabla^2 u\right)\nabla^2 u\right] \\ u(0, x, y) = u_0(x, y) \end{cases}, \tag{2.23}$$

where $g(s) = \frac{f'(s)}{s}$ [39]. The You–Kaveh filter provides effective denoising results and overcomes the staircase effect but it may also generate multiplicative (speckle) noise. For this reason, a despeckling procedure is also provided in [39].

Numerous fourth-order diffusion-based techniques improving the Y-K model have been developed in the last years. GVC-based fourth-order anisotropic diffusion filtering model introduced by Wang et al. inserts the Gradient Vector Convolution (GVC) field into You–Kaveh scheme [40], as following:

$$\frac{\partial u}{\partial t} = -\Delta(c_1 \Delta u - c_2 V_{GVC} \cdot \mathrm{N}) \tag{2.24}$$

where $c_1(|\nabla u|) = \frac{1}{1+\left(\frac{|\nabla u|}{k_1}\right)^2}$, $c_2(|\nabla u|) = m|\nabla u| e^{-\left(\frac{|\nabla u|}{k_2}\right)^2}$, $m \in (0, 1]$ and $\mathrm{N} = \frac{\nabla u}{|\nabla u|}$.

This GVC-based restoration model provides a better boundary and texture preservation, due to the outstanding edge detection ability of the GVC field. This fourth-order anisotropic diffusion filter provides a more effective AWGN reduction, illustrated by its higher PSNR scores. It also improved the numerical stability [40].

An improvement of the You–Kaveh model is based on an adaptive relaxation method and a discontinuity treatment of the edges [41]. Another well-known nonlinear fourth-order PDE-based restoration approach is the variational LLT model developed by Lysaker et al. [42]. They used 2 different functions to measure the oscillations of the noisy image data. LLT denoising is based on the next energy minimization:

$$u = \arg\min_u \int_\Omega \left(\left|D^2(u)\right| + \frac{\lambda}{2}(u - u_0)^2 \right) d\Omega \tag{2.25}$$

where $\lambda \geq 0$ and $\left|D^2(u)\right| = \begin{cases} |u_{xx}| + |u_{yy}|, \; or \\ \left(u_{xx}^2 + u_{xy}^2 + u_{yx}^2 + u_{yy}^2\right)^{1/2} \end{cases}$. It outperforms the Y-K model for a proper λ value. It provided very good detail-preserving noise removal results on the medical images [42].

An improved nonlinear fourth-order diffusion-based filter was modeled by T. Chan by inserting a fourth-order diffusion term to the Euler–Lagrange equation of TV Denoising [43]. It alleviate considerably the blocky effect and preserve well the features. The fourth-order PDE models with *piecewise planarity conditions* (PPCs) were introduced by Kim and Lim [44]. With these conditions the unconstrained 4th-order variational schemes in continuum converge to a piecewise planar image.

While nonlinear fourth-order diffusion-based filters reduce the AWGN successfully and overcome the staircase effect, they also have their own drawbacks. Besides producing some multiplicative noise, they may generate image over-filtering. Since fourth-order diffusion damps the high frequency image components much faster, it could over-smooth the step boundaries. Some restoration approaches that combine second and fourth order PDE-based filters, and enjoy the advantages of both types of nonlinear PDE models, have been introduced for this reason.

Thus, some restoration frameworks use LLT in combination with second-order anisotropic diffusion filters like the Perona–Malik scheme. Another compound diffusion-based restoration solution combines fourth-order LLT to a second-order TV filter [45]. A restoration method that combines a second-order P-M type model to a fourth-order You–Kaveh type scheme was introduced by Liu and Xiang [46]. They approximate the models using finite differences, then, a convex combination of the two stable numerical approximation algorithms is performed. The obtained approach removes successfully the AWGN, while avoiding all the undesired effects. Another convex combination of two nonlinear PDE-based models of different orders, which outperforms each of its two PDE-based components, has the form $w = \theta u + (1-\theta)v, \theta \in [0, 1]$ [47], where $u(0, x, y) = u_0(x, y)$, $v(0, x, y) = v_0(x, y)$, $|D^2 v| = \left(v_{xx}^2 + v_{xy}^2 + v_{yx}^2 + v_{yy}^2\right)^{1/2}$ and

$$\begin{cases} u_t = \nabla \cdot \left(\frac{\nabla u}{|\nabla u|}\right) - \lambda_1 (u - u_0) \\ and \\ v_t = -\left(\frac{v_{xx}}{|D^2 v|}\right)_{xx} - \left(\frac{v_{xy}}{|D^2 v|}\right)_{xy} - \left(\frac{v_{yx}}{|D^2 v|}\right)_{yx} - \left(\frac{v_{yy}}{|D^2 v|}\right)_{yy} - \lambda_2 (v - v_0) \end{cases} \tag{2.26}$$

Other hybrid denoising techniques mix nonlinear fourth-order PDE models of different types. Such a compound approach combines a mean curvature motion-based scheme to a nonlinear fourth-order PDE [48], having the form:

$$\begin{cases} u_t = -|\nabla u|^{\alpha} \Delta \cdot \left(\frac{\Delta u}{|\Delta u|^{1+\varpi}}\right) - \beta (u - u_0), (x, y) \in \Omega \\ u(x, y, 0) = u_0(x, y), (x, y) \in \Omega \\ \frac{\partial u}{\partial \vec{n}} = 0, (x, y) \in \partial \Omega \end{cases} \tag{2.27}$$

with $\alpha, \beta, \omega \geq 0$. It overcomes the blocky artifacts and removes speckle noise, too.

We have also developed many nonlinear parabolic and hyperbolic fourth-order PDE-based denoising techniques in the last years, which are surveyed in [22]. Let us mention here our hyperbolic PDE-based filter that is based on the model [49]:

$$\begin{cases} \xi \frac{\partial^2 u}{\partial t^2} + \gamma^2 \frac{\partial u}{\partial t} + \Delta\left(\varphi_u(\|\nabla u\|) \cdot \nabla^2 u\right) + \lambda (u - u_0) = 0 \\ u(0, x, y) = u_0(x, y) \\ u_t(0, x, y) = u_1(x, y) \\ u(t, x, y) = 0, \quad \forall t \geq 0, \quad (x, y) \in \partial \Omega \end{cases} \quad (x, y) \in \Omega \tag{2.28}$$

where $\xi, \gamma, \lambda \in (0, 1]$, u_1 is a velocity modification of the observed image, $\varphi_u(s) = \delta \sqrt{\frac{\psi(u)}{\alpha \log 10\left(s^2 + \psi(u)\right)^k + \beta}}$, $\psi(u) = |\eta \mu(\|\nabla u\|) - \varepsilon|, \alpha, \eta \in (0, 1.5), k, \beta, \varepsilon \in [2, 5)$.

This hyperbolic fourth-order PDE-based denoising scheme outperforms both the second and fourth order diffusion-based filters, including P-M and Y-K models,

Table 2.1 Method comparison: average PSNR scores of several AWGN filters

Method	This approach (dB)	P-M 1 (dB)	P-M 2 (dB)	TV denoising (dB)	You–Kaveh (dB)	Gaussian filter (dB)
PSNR	28.34	26.23	24.91	21.88	27.19	21.37

providing more effective noise removal results, sharpening the boundaries, executing faster and overcoming the side-effects [49].

We have also proposed PDE-based restoration solutions combining succesfully second and fourth-order diffusions. Such a variational denoising scheme minimizes

$$J(u) = \int_{\Omega} \left(\varphi_1(\|\nabla u\|) + \varphi_2(\nabla^2 u) + \frac{\lambda}{2}(u - u_0)^2 \right) d\Omega \tag{2.29}$$

where φ_1 and φ_2 are 2 regularizers, $\lambda \in (0, 1)$. It leads to the fourth-order PDE model

$$\begin{cases} \frac{\partial u}{\partial t} = div\left(\psi_1^u(\|\nabla u\|)\nabla u\right) - \nabla^2\left(\psi_2^u\left(\left|\nabla^2 u\right|\right)\nabla^2 u\right) - \lambda(u - u_0) \\ u(0, x, y) = u_0 \\ u(t, x, y) = 0, \quad on\ \partial\Omega \\ \dfrac{\partial u}{\partial \vec{n}} = 0 \end{cases} \tag{2.30}$$

with $\psi_1^u, \psi_2^u : [0, \infty) \to (0, \infty)$, $\psi_1^u(s) = \frac{1}{s}\frac{\partial \varphi_1(s)}{\partial s}$, $\psi_2^u(s) = \frac{1}{s}\frac{\partial \varphi_2(s)}{\partial s}$ [50]. It outperforms both classic and the second and fourth-order PDE-based filters, as illustrated by method comparison results and example in Table 2.1 and Fig. 2.2.

2.2 Nonlinear PDE-Based Photon-Limited Image Restoration Solutions

While AWGN represents the most common type of noise that affects the digital images during the capturing process, there are also other noises that may be generated by the acquisition devices. One of them is the *quantum*, or *shot, noise*, which is also called *Poisson noise* since it is characterized by the Poisson distribution: $P(n) = \frac{e^{-\mu}\mu^n}{n!}, n \geq 0$ [51]. While AWGN is a signal-independent noise, the Poisson noise is a signal-dependent one, produced by the imaging instruments based on photon-counting devices. Due to the inherently random nature of the photon emission and detection, the signal-dependent errors of the image acquisition mechanisms generate the *photon-limited images*.

Many photon-limited image restoration solutions have been provided in the last decades. Some conventional filters, like the 2D median filter [2], reduce the Poisson noise but they may affect the image details. More effective quantum noise

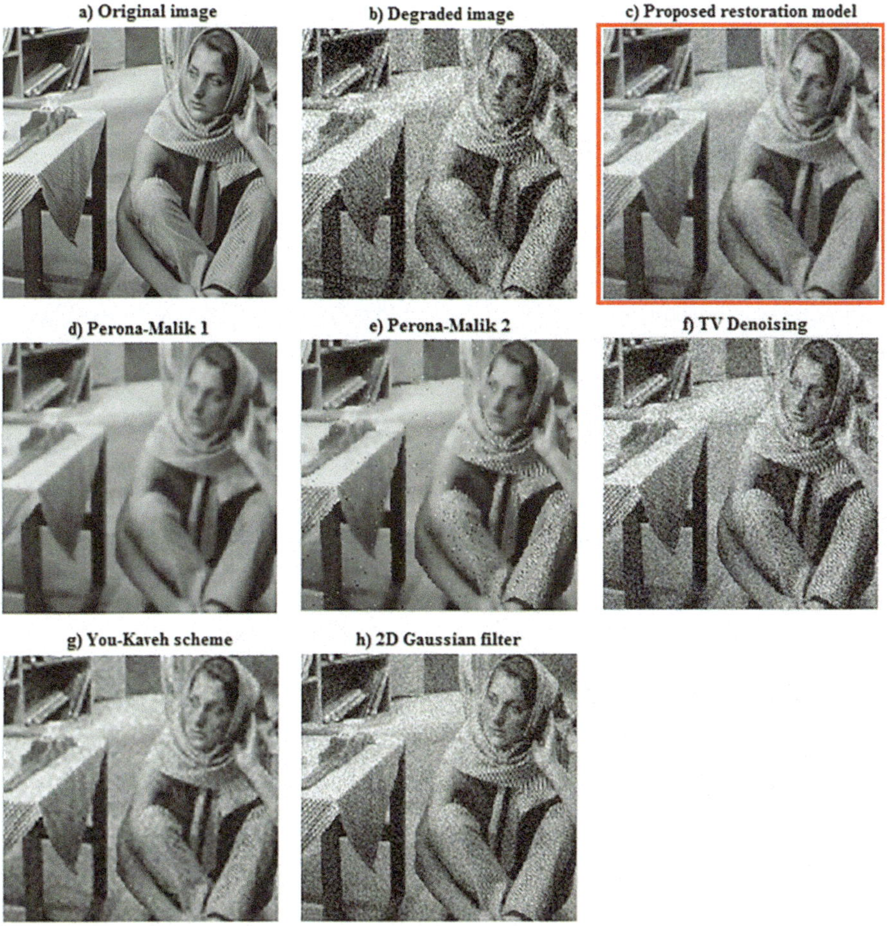

Fig. 2.2 Image filtered by several PDE and non-PDE filters

removal approaches include the Non-Local Mean—NLM Poisson filter [52], Poisson Reducing Bilateral Filter—PRBF [53], Multi Scale Variance Stabilizing Transform (MS-VST) [54], Minimum Description Length (MDL)-based method [55], wavelet, ridgelet and curvelet-based schemes [56].

Partial differential equation (PDE)-based filters represent also very effective photon-limited image denoising tools. Many nonlinear PDE-based quantum noise removal models were derived from PDE-based additive Gaussian noise filtering schemes. Thus, total variation regularization-based quantum denoising models were obtained adapting the TV-based AWGN filtering schemes to the Poisson distribution, by inserting a data-fidelity term that is suitable for shot noise [57–59].

Such a variational PDE photon-limited image restoration approach was introduced by Le et al. [57]. It applies the next energy functional minimization:

$$u_{rest} = \arg\min_{u \in BV(\Omega)} \int_{\Omega} (|\nabla u| + \lambda(u - u_0 \log u))d\Omega \tag{2.31}$$

where u_0 is the photon-limited observation, $BV(\Omega)$ is the space of the bounded variation functions on the image domain $\Omega \subseteq R^2$ and λ stands for a penalty term. The TV-based model (2.31) generates a nonlinear parabolic second-order PDE-based filter that achieves a good trade-off between noise reduction and feature preservation, but could produce blocky effects [57]. For this reason, several fourth-order PDE denoising models for shot noise have been developed. A energy-based restoration technique was introduced by Zhou and Li in [58]. It denoises properly the photon-limited images using the energy minimization:

$$u_{\min} = \min_{u} \left\{ E(u) = \int_{\Omega} \left(\left| D^2 u \right| + \lambda(u - u_0 \log u) \right) d\Omega \right\} \tag{2.32}$$

where Du is the distributional derivative of u. Then, a nonlinear fourth-order PDE is derived from (2.32). The obtained model is solved numerically applying the alternating minimization (AM) algorithm [58]. The fourth-order PDE-based approach removes successfully the Poisson noise and overcomes the staircase effect. It outperforms the second-order PDE-based quantum denoising models, as illustrated by the method comparison results described in Table 2.2 and Fig. 2.3, where these 2 variational approaches given by (2.31) and (2.32) are compared [58].

Another energy-based quantum noise removal technique was proposed by Sawatzky et al. [59]. Their PDE-based filter uses the total variation minimization:

$$u_{\min} = \arg\min_{u \in BV(\Omega)} \left\{ \frac{1}{2} \int_{\Omega} \left(\frac{(u - u_0)^2}{u_0} + \alpha |u|_{BV} \right) d\Omega \right\} \tag{2.33}$$

where $u_{\min}$ is the restored image, u_0 is the photon-limited image and $\alpha \geq 0$. This variational method provides an effective Poisson noise removal. A filtering example is described in Fig. 2.4. The original image in (a) is then contaminated by a high amount of signal-dependent noise in (b), its restored version being depicted in (c).

Other variational restoration approaches, such as the *fractional-order total variation-based Poisson denoising* (FOTV) model introduced by Chowdhury et al. in 2020 [60], use higher order total variations. So, the FOTV-regularized denoising technique is characterized by the minimization problem:

Table 2.2 Results obtained by the two TV-based shot denoising models

Method	Iterations	Running time (s)	Relative error
2nd order PDE model	450	17:21	0.0541
4th order PDE model	250	10:04	0.0530

Fig. 2.3 Results of 2nd and 4th order PDE models on *Panda* image

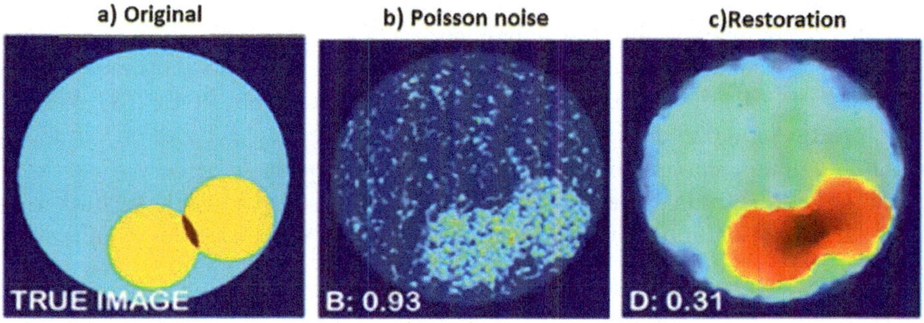

Fig. 2.4 Variational PDE-based Poisson denoising example

$$\min_{u \in BV^{\alpha}(\Omega)} TV^{\alpha}(u) + \beta \int_{\Omega} (u - u_0 \log u) d\Omega \tag{2.34}$$

where $TV^{\alpha}(u)$ stands for the α-order total variation and the fractional order $\alpha > 0$ [60]. The nonlinear PDE model derived from this variational scheme is well-posed under some certain assumptions. The existence and uniqueness of a variational solution for this PDE is demonstrated for $\beta > 0$ in [60].

This variational approach outperforms some well-known non-PDE based Poisson noise reduction techniques [60]. Some noise removal results are displayed in Fig. 2.5, where the original image is represented first, the second one is the photon-limited

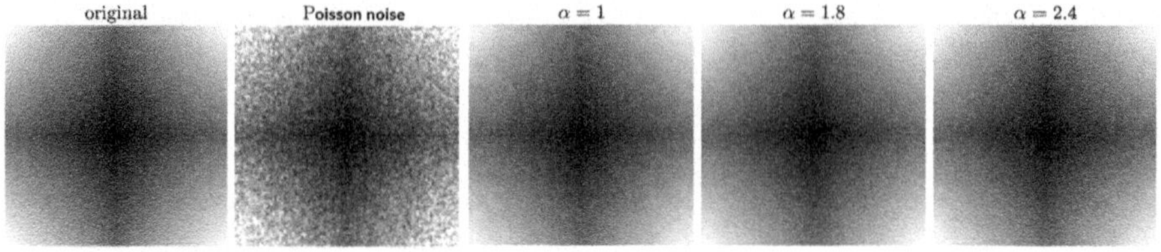

Fig. 2.5 FOTV-based denoising results

image and the next three images display the FOTV-based restoration results for several fractional order parameter values.

The following total bounded variation-based Poissonian images restoration model was proposed by Liu and Huang [61]:

$$\min_{u \in BV(\Omega), \log Ku \in L^1(\Omega)} \|\nabla u\|_1 + \frac{\alpha}{2}\|u\|_2^2 + \beta \int_\Omega (Ku - u_0 \log Ku)d\Omega \tag{2.35}$$

where $\alpha \geq 0, \beta > 0$ and K is a bounded linear operator representing the convolution kernel. Its optimal solution was determined recursively using an extended split Bregman. This restoration model preserves the boundaries while removing the Poisson noise, but may cause blocky effects in the flat regions [61].

Other quantum noise removal techniques use nonlinear fourth-order PDE models solved by the alternating minimization algorithm [62], PDE-based models based on Euler's Elastica Regularization [63] and Hamilton–Jacobi PDEs [64]. We also developed several photon-limited image restoration techniques based on variational and non-variational, parabolic and hyperbolic PDE models.

In [65] we disseminated an anisotropic diffusion-based quantum noise filter using the next second-order nonlinear parabolic PDE model with boundary conditions:

$$\begin{cases} \frac{\partial u}{\partial t} - \delta(\|\nabla u\|)\nabla \cdot (\psi(\|\nabla u\|)\nabla u) + \lambda \frac{u-u_0}{u} = 0 \\ u(x, y, 0) = u_0(x, y), \quad \forall (x, y) \in \Omega \\ \frac{\partial u}{\partial \vec{n}}(x, y) = 0, \quad \forall (x, y) \in \partial\Omega \end{cases} \tag{2.36}$$

where $\lambda \in [0, 1)$ and the proposed diffusivity function is

$$\psi : [0, \infty) \to [0, \infty), \psi(s) = \left(\frac{|\mu(\|\nabla u\|) + \alpha t|}{|\beta \ln(\mu(\|\nabla u\|) + \alpha t) + \eta s^3|} \right)^{\frac{1}{2}} \tag{2.37}$$

where $\eta \in [0, 3)$, $\alpha, \beta \in [0, 1)$ and

$$\delta : [0, \infty) \to [0, \infty), \delta(s) = \frac{\sqrt[3]{\gamma s^2 + \zeta}}{\xi} \tag{2.38}$$

with $\gamma, \xi, \zeta \in [1, 4)$ [65]. The component $\lambda \frac{u-u_0}{u}$ is related to the Poisson distribution. Unlike other PDE-based quantum noise removal schemes, this diffusion model is non-variational, since it cannot be derived from any energy functional minimization. Given Neumann boundary conditions taken in (2.36), this nonlinear second-order PDE model is also well-posed, admitting a unique weak solution that is determined numerically using a finite difference-based approximation algorithm [65].

It uses a grid of space size h and time step Δt, which quantizes the time and space coordinates as $x = ih, y = jh, t = n\Delta t, i \in \{1, \ldots, I\}, j \in \{1, \ldots, J\}, n \in \{0, \ldots, N\}$, for a $[Ih \times Jh]$ image. Then, $\frac{\partial u}{\partial t} + \lambda \frac{u-u_0}{u}$ is approximated as $\frac{u_{i,j}^{n+\Delta t} - u_{i,j}^{n}}{\Delta t} + \lambda \frac{\left(u_{i,j}^{n} - u_{i,j}^{0}\right)}{u_{i,j}^{n}}$. One computes $\delta_{i,j} = \delta\left(\left\|\nabla u_{i,j}\right\|\right)$, with $\left\|\nabla u_{i,j}\right\| \approx \sqrt{\left(\frac{u_{i+h,j}^{n} - u_{i-h,j}^{n}}{2h}\right)^2 + \left(\frac{u_{i,j+h}^{n} - u_{i,j-h}^{n}}{2h}\right)^2}$. The divergence component, $\nabla \cdot (\psi(\|\nabla u\|)\nabla u)$, could be discretized as

$$\frac{\psi\left(\left\|u_{i+h,j}^{n} - u_{i,j}^{n}\right\|\right)\left(u_{i+h,j}^{n} - u_{i,j}^{n}\right) + \psi\left(\left\|u_{i-h,j}^{n} - u_{i,j}^{n}\right\|\right)\left(u_{i-h,j}^{n} - u_{i,j}^{n}\right) + \psi\left(\left\|u_{i,j+h}^{n} - u_{i,j}^{n}\right\|\right)\left(u_{i,j+h}^{n} - u_{i,j}^{n}\right) + \psi\left(\left\|u_{i,j-h}^{n} - u_{i,j}^{n}\right\|\right)\left(u_{i,j-h}^{n} - u_{i,j}^{n}\right)}{4}.$$

If one consider $h = \Delta t = 1$, one obtains the explicit numerical approximation scheme

$$\begin{aligned} u_{i,j}^{n+1} = {} & \frac{\delta_{i,j}}{4}\Big[\psi\left(\left\|u_{i+1,j}^{n} - u_{i,j}^{n}\right\|\right)\left(u_{i+1,j}^{n} - u_{i,j}^{n}\right) + \psi\left(\left\|u_{i-1,j}^{n} - u_{i,j}^{n}\right\|\right)\left(u_{i-1,j}^{n} - u_{i-1,j}^{n}\right) \\ & + \psi\left(\left\|u_{i,j+1}^{n} - u_{i,j}^{n}\right\|\right)\left(u_{i,j+1}^{n} - u_{i,j}^{n}\right) + \psi\left(\left\|u_{i,j-1}^{n} - u_{i,j}^{n}\right\|\right)\left(u_{i,j-1}^{n} - u_{i,j}^{n}\right)\Big] \\ & + u_{i,j}^{n} - \lambda \frac{u_{i,j}^{n} - u_{i,j}^{0}}{u_{i,j}^{n}}, \quad \forall n \in \{0, \ldots, N\} \end{aligned} \tag{2.39}$$

with discrete Neumann boundary conditions $u_{2,j}^{n} - u_{1,j}^{n} = 0$, $u_{J,j}^{n} - u_{J-1,j}^{n} = 0$, $u_{i,2}^{n} - u_{i,1}^{n} = 0$, $u_{i,I}^{n} - u_{i,I-1}^{n} = 0$. The scheme (2.39) is consistent to the PDE model and converges fast to its solution representing the denoised image. However, its number of iterations, N, depends on the image size and amount of shot noise. This denoising algorithm removes success-fully this type of noise and overcomes the blurring and the staircase artifacts, while preserving the essential image details. The image datasets used in our simulations were the volumes of the USC-SIPI database [65]. It outperforms conventional filters approaches, like 2D Median filter, and the TV-based models adapted for Poisson noise, as shown by the method comparison results in Table 2.3 and Fig. 2.6.

Another non-variational second-order PDE-based Poissonian image restoration approach introduced by us was based on the following hyperbolic model [66]:

Table 2.3 The average PSNR values of several filters

Restoration technique	Average PSNR value (dB)
The proposed AD-based method	34.2536
Median 2D filter	29.7369
NLM filter	31.3794
Bilateral 2D filter	29.8467
TV model for Poisson noise	33.2542

Fig. 2.6 Photon-limited image restoration output of several filters

$$\begin{cases} \lambda \frac{\partial^2 u}{\partial t^2} + \xi \frac{\partial u}{\partial t} - \beta\varphi\left(\left\|\nabla^2 u\right\|\right) div(\psi(\|\nabla u\|)\nabla u) + \alpha\left(\frac{u-u_0}{|u|+\delta}\right) = 0 \\ u(x, y, 0) = u_0(x, y), \quad \forall(x,y) \in \Omega \subseteq R^2 \\ u_t(x, y, 0) = u_1(x, y), \quad \forall(x,y) \in \Omega \\ u(x, y, t) = 0, \quad \forall(x,y) \in \partial\Omega \\ \frac{\partial u}{\partial \vec{n}}(x, y, t) = 0, \quad \forall(x,y) \in \partial\Omega \end{cases} \tag{2.40}$$

where $\alpha, \beta, \lambda, \xi, \delta \in (0, 1]$, $u_0 \in L^2(\Omega)$ and

$$\psi : [0, \infty) \rightarrow [0, \infty), \psi(s) = \left(\frac{\eta}{\left|\zeta s^k + \gamma \log 10(\eta)\right|}\right)^{\frac{1}{k-1}} \tag{2.41}$$

where $\zeta \in (0, 1]$, $\gamma \in (0.5, 1]$, $k \geq 3$, $\eta \geq 20$ and $\varphi(s) = \frac{1}{\varepsilon}\left(\upsilon s^{r-1} + \varsigma\right)^{\frac{1}{r}}$.

The term $div(\psi(\|\nabla u\|)\nabla u)$ performs the diffusion along the edges, so overcoming the blurring, and $\varphi\left(\left\|\nabla^2 u\right\|\right)$ controls the speed of the diffusion process. This filter sharpens the boundaries and enhances the image details, given its second-time derivative, $\frac{\partial^2 u}{\partial t^2}$, that provides the hyperbolic character of the nonlinear PDE model. The PDE model (2.40) is well-posed, since it exists a unique variational solution for it. A numerical approximation algorithm that solves it model and converges to that solution is constructed using the finite difference method [66]. A grid that is similar to that in the previous case is used here and the explicit iterative scheme is obtained:

$$u_{i,j}^{n+1}\left(\frac{2\lambda+\xi}{2}\right)=u_{i,j}^{n}\left(2\lambda-\frac{\alpha}{\left|u_{i,j}^{n}\right|+\delta}\right)+u_{i,j}^{n-1}\left(\frac{\xi-2\lambda}{2}\right)+u_{i,j}^{0}\frac{\alpha}{\left|u_{i,j}^{n}\right|+\delta}$$
$$+\beta\frac{\varphi\left(u_{i+1,j}^{n}+u_{i-1,j}^{n}+u_{i,j+1}^{n}+u_{i,j-1}^{n}-4u_{i,j}\right)}{4}$$
$$\begin{bmatrix}\psi\left(\left\|u_{i+1,j}^{n}-u_{i,j}^{n}\right\|\right)\left(u_{i+1,j}^{n}-u_{i,j}^{n}\right)+\psi\left(\left\|u_{i-1,j}^{n}-u_{i,j}^{n}\right\|\right)\left(u_{i-1,j}^{n}-u_{i,j}^{n}\right)\\+\psi\left(\left\|u_{i,j+1}^{n}-u_{i,j}^{n}\right\|\right)\left(u_{i,j+1}^{n}-u_{i,j}^{n}\right)+\psi\left(\left\|u_{i,j-1}^{n}-u_{i,j}^{n}\right\|\right)\left(u_{i,j-1}^{n}-u_{i,j}^{n}\right)\end{bmatrix} \tag{2.42}$$

where $n \in \{0, \ldots, N\}, i \in \{1, \ldots, I\}, j \in \{1, \ldots, J\}$. This numerical approximation algorithm is consistent to the PDE model and converges fast to its weak solution [66].

This Poisson denoising framework was tested successfully on hundreds of photon-limited medical images. Our restoration experiments were performed on the microscope image collections of the Public Health Image Library (PHIL) of the Centers for Disease Control and Prevention [66]. The iterative numerical algorithm filters properly the shot noise, overcomes the undesired effects, such as image blurring and staircasing, and preserves and even enhances essential features such as edges, and corners. Since it converges fast to the solution of the PDE model, its number of iterations N is quite low. Also, this filtering technique outperforms both non-PDE denoising models, like 2D median, bilateral and NLM filters, and some second-order variational PDE-based filters, providing better performance measures values, as shown by the method comparison results in Table 2.4 and Fig. 2.7 [66].

Another second-order PDE-based photon-limited medical image restoration technique, was proposed by us in [67]. Unlike the previously described non-variational denoising method, this one is based on a variational model given by the minimization $u_{\min} = \arg\min_{u \in BV(\Omega)} F(u)$, where

Table 2.4 Method comparison: average PSNR, MSE and SNR values

Technique	Average PSNR (dB)	Average MSE	Average SNR
This approach	33.3697	29.9301	33.0576
Median filter	28.8361	85.0093	28.2003
Bilateral 2D filter	29.2061	78.0666	28.6433
TV-based filter	30.8756	60.5191	30.3135
NLM filter	31.7068	43.8936	31.3238

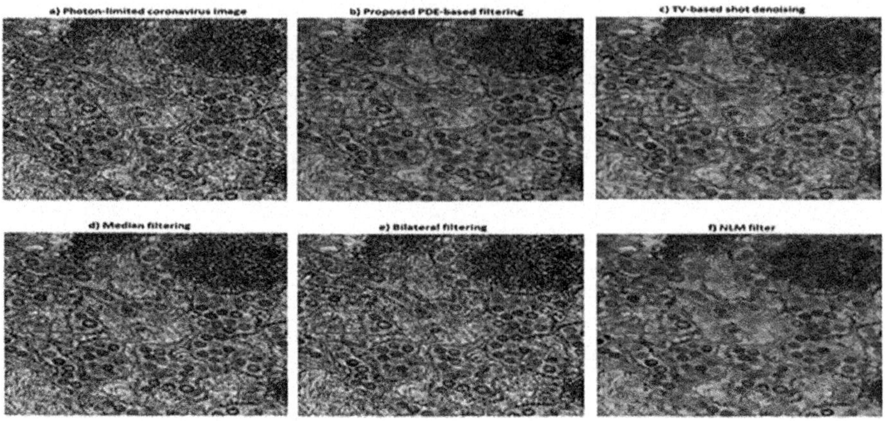

Fig. 2.7 Photon-limited electron microscope image restored by various methods

$$F(u) = \int_{\Omega} \left(\alpha\psi\left(\left|\nabla^2 u\right|\right) - \xi(u - u_0 \ln u)\right) d\Omega \tag{2.43}$$

where $\alpha, \xi \in (0, 1]$, $u_0 \in L^2(\Omega)$ is the photon-limited image observation and

$$\psi(s) = \int_0^s \tau\lambda \left(\frac{\beta}{\nu\left|\ln\left(s^k + \beta\right)^3\right| + \gamma} \right)^{1/2} d\tau \tag{2.44}$$

with $\nu, \lambda, \gamma \in (0, 5]$, $\beta \geq 25$ and $k \geq 2$.

We note $E(x, y, u, u_x, u_y) = \alpha\psi\left(\left|\nabla^2 u(x, y)\right|\right) + \xi(u(x, y) - u_0(x, y) \ln u(x, y))$ and apply the next Euler–Lagrange equation corresponding to the variational model:

$$\frac{\partial E}{\partial u} - \frac{\partial}{\partial x}\frac{\partial E}{\partial u_x} - \frac{\partial}{\partial y}\frac{\partial E}{\partial u_y} = 0 \tag{2.45}$$

which leads to

$$\alpha\Delta\left(\frac{(\psi)'(|\Delta u|)}{|\Delta u|}\Delta u\right) - \frac{\xi(u - u_0)}{u} = 0 \tag{2.46}$$

By applying the steepest gradient descent method, one obtains the following parabolic nonlinear fourth-order PDE-based model [67]:

$$\begin{cases} \frac{\partial u}{\partial t} = -\alpha\nabla^2\left(\varphi(\|\Delta u\|)\nabla^2 u\right) + \frac{\xi(u-u_0)}{u}, & \forall (x,y) \in \Omega \\ u(x, y, 0) = u_0(x, y), & \forall (x,y) \in \Omega \subseteq R^2 \\ \frac{\partial u}{\partial \vec{n}}(x, y, t) = 0, & \forall (x,y) \in \partial\Omega \end{cases} \tag{2.47}$$

where $\varphi(s) = \frac{1}{s}\frac{\partial\psi(s)}{\partial s} = \lambda\sqrt{\frac{\beta}{\nu\left|\ln\left(s^k+\beta\right)^3\right|+\gamma}}$ [67].

This nonlinear diffusion model is well-posed and its numerical solution is determined by applying iteratively this finite difference-based algorithm [67]:

$$u_{i,j}^{n+1} = u_{i,j}^{n}\left(1 - \frac{\xi}{u_{i,j}^{n}}\right) + \alpha\left(\eta_{i+1,j}^{n} - \eta_{i-1,j}^{n} - \eta_{i,j+1}^{n} - \eta_{i,j-1}^{n} + 4\eta_{i,j}^{n}\right) + u_{i,j}^{0}\frac{\xi}{u_{i,j}^{n}} \tag{2.48}$$

where $\eta_{i,j}^{n} = \varphi\left(\left|\nabla^2 u_{i,j}^{n}\right|\right)\Delta u_{i,j}^{n}$. This numerical scheme is consistent to the PDE model, removes effectively the shot noise, overcomes side-effects, such as blurring and staircasing, and conserves the essential image details. Some method comparison results based on our experiments performed on voluminous medical imaging datasets, like Brain Tumor Progression collection of Cancer Imaging Archive (TCIA), and illustrating our filter's effectiveness, are provided in Table 2.5 and Fig. 2.8.

Table 2.5 The average PSNR and SSIM values of several filters

Restoration technique	Average PSNR (dB)	Average SSIM
The proposed technique	32.5082	1.2496
2D median filter	28.8973	1.1121
TV-based Poisson denoising	30.8456	1.1872
NLM filter	29.1354	1.1205
Bilateral 2D filter	31.0938	1.2003

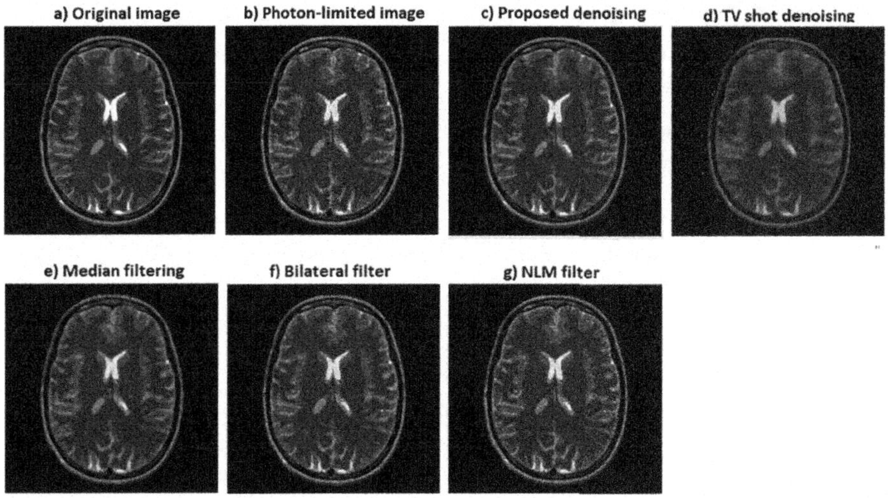

Fig. 2.8 Photon-limited medical image restored by several filters

A hyperbolic fourth-order PDE-based shot denoising technique was disseminated in [68]. It is based on this partial differential equation with boundary conditions:

$$\begin{cases} \alpha \frac{\partial^2 u}{\partial t^2} + \beta \frac{\partial u}{\partial t} + \lambda \nabla^2 \left(\varphi(\|\nabla u\|) \nabla^2 u \right) + \varepsilon \left(\frac{u - u_0}{|u| + \eta} \right) = 0 \\ u(x, y, 0) = u_0(x, y), \quad \forall (x,y) \in \Omega \subseteq R^2 \\ u_t(x, y, 0) = u_1(x, y), \quad \forall (x,y) \in \Omega \\ u(x, y, t) = 0, \quad \forall (x,y) \in \partial\Omega \\ \frac{\partial u}{\partial \vec{n}}(x, y, t) = 0, \quad \forall (x,y) \in \partial\Omega \end{cases} \tag{2.49}$$

where $\alpha, \beta, \lambda, \varepsilon, \eta \in (0, 1]$, $\varphi(s) = \sqrt[k-1]{\frac{\delta(x,y,t)}{|\gamma s^k + \xi \log 10(\delta(x,y,t))|}}$, $\xi \in (0.5, 1]$, $\gamma \in (0, 1] k \geq 3$ and conductance parameter δ depends on the image coordinates and statistics.

The well-posedness of this PDE restoration model is mathematically treated in [68]. We set $H = L^2(\Omega)$, $V = H^2(\Omega) \cap H_0^1(\Omega)$ and note that $V \subset H$ with dense and compact embedding. Also, the Lipschitz property of φ is important to our investigation [69]. Function $u : [0, T] \times \Omega \rightarrow R$ is a variational solution to (2.2) if next conditions hold

$$\begin{cases} u, u_t \in L^\infty(0, T; V), u_{tt} \in L^\infty(0, T; H) \\ u(x, y, 0) = u_0(x, y) \\ \int\limits_\Omega (\alpha u_{tt}(x, y.t) + \beta u_t(x, y, t)) v(x, y, t) dx dy \\ +\lambda \int\limits_\Omega \varphi(\|\nabla u\|)(x, y, t) \Delta u(x, y, t) \Delta v(x, y, t) dx dy \\ +\varepsilon \int\limits_\Omega \frac{u(x,y,t) - u_0(x,y)}{|u(x,y,t)| + \eta} v(x, y, t) dx dy = 0, \quad \forall v \in V, \quad t \in [0, T] \end{cases} \tag{2.50}$$

To demonstrate this model's validity, one constructs the sequence $\{u_n\} \subset L^\infty(0, T; V)$, where $(u_n)_t \subset L^\infty(0, T; V)$, $(u_n)_{tt} \subset L^\infty(0, T; H)$, defined by:

$$\begin{cases} \alpha (u_n)_{tt} + \beta (u_n)_t + \lambda \Delta (\varphi(\|\nabla u_{n-1}\|) \Delta u_n) + \varepsilon \frac{u_n - u_0}{|u_n| + \eta} = 0 \\ u_n(x, y, t) = 0, \frac{\partial u_n}{\partial n}(x, y, t) = 0 \text{ on } (0, T) \times \partial\Omega \\ u_n(x, y, t) = u_0, (u_n)_t(x, y, 0) = u_1 \text{ on } \Omega \end{cases} \tag{2.51}$$

One applies Theorem 1.1 given in [69] and it follows that for any u_{n-1} there is a unique weak solution u_n to (2.49). Then, one can prove the convergence of $\{u_n\}$ to a weak solution u for the PDE-based model (2.1), by demonstrating just that $\{u_n\}$ and $\{(u_n)_t\}$ are bounded in $L^2(0, T; V)$ and $\{(u_n)_{tt}\}$ is bounded in $L^2(0, T; H)$, and so $\{u_n\}$ and $\{(u_n)_t\}$ are compact in $L^2(0, T; H)$. If one multiplies the equation by u_n, and then integrates on $[0, t] \times \Omega$, one obtains after some computing:

$$\alpha \int_{\Omega} u_n(x, y, t)(u_n)_t(x, y, t)dxdy + \frac{\beta}{2} \int_{\Omega} u_n^2(x, y, t)dxdy$$
$$+ \lambda \int_0^t \int_{\Omega} \varphi(\|\nabla u_{n-1}\|)|\Delta u_n|^2 dsdxdy$$
$$= \frac{\beta}{2} \int_{\Omega} |u_0|^2 dxdy + \alpha \int_{\Omega} u_1 u_0 dxdy$$
$$+ \alpha \int_0^t \int_{\Omega} (u_n)_t^2 dsdxdy - \varepsilon \int_0^t \int_{\Omega} \frac{u_n - u_0}{|u_n| + \eta} u_n dsdxdy \tag{2.52}$$

and therefore

$$\beta \int_{\Omega} u_n(x, y, t)dxdy + \int_0^t \int_{\Omega} |\Delta u_n|^2 dsdxdy \le c \int_0^t \int_{\Omega} (u_n)_t^2 dsdxdy + c, \quad \forall t \in [0, T] \tag{2.53}$$

where $c > 0$ is independent of n. Next, we multiply (2.49) by $(u_n)_t$ and then integrate it, to achieve the next estimate:

$$\frac{\alpha}{2} \int_{\Omega} (u_n)_t^2(x, y, t)(u_n)_t dxdy + \beta \int_{\Omega} u_n^2(x, y, t)dxdy$$
$$+ \frac{\lambda}{2} \int_0^t \int_{\Omega} \varphi(\|\nabla u_{n-1}\|)|\Delta u_n|_s^2 dxdyds$$
$$= -\varepsilon \int_0^t \int_{\Omega} \frac{u_n - u_0}{|u_n| + \eta} (u_n)_t dsdxdy + \frac{\alpha}{2} \int_{\Omega} u_1^2 dxdy \tag{2.54}$$

Then, multiplying by Δu_n and $\Delta(u_n)_t$ and using (2.51) and (2.52), one obtains:

$$\int_{\Omega} \left((u_n)_t^2 + u_n^2\right)dxdy + \int_0^t \int_{\Omega} |\Delta u_n|^2 dsdxds$$
$$+ \int_0^t \int_{\Omega} |\nabla(u_n)_t|^2 dsdxdy \le c, \quad \forall t \in [0, T] \tag{2.55}$$

Table 2.6 PSNR values several Poisson noise removal approaches

Restoration model	Average PSNR (dB)
The proposed hyperbolic PDE model	34.3541
Median filter	27.9572
TV model for Poisson noise	31.4754
Bilateral 2D filter	29.8968
NLM filter	28.4369

as claimed. The solution of this nonlinear hyperbolic PDE model is determined numerically applying the finite difference-based approximation algorithm in [68]:

$$u_{i,j}^{n+1} = u_{i,j}^{n}\left(\frac{4\alpha\left(\left|u_{i,j}^{n}\right| + \eta\right) - 2\varepsilon}{(2\alpha+\beta)\left(\left|u_{i,j}^{n}\right| + \eta\right)}\right) + u_{i,j}^{n-1}\left(\frac{\beta - 2\alpha}{\beta + 2\alpha}\right) + u_{i,j}^{0}\frac{2\varepsilon}{(2\alpha+\beta)\left(\left|u_{i,j}^{n}\right| + \eta\right)}$$
$$- \frac{2\lambda\left(\varphi^{i+1,j} + \varphi^{i-1,j} + \varphi^{i,j+1} + \varphi^{i,j-1} - 4\varphi^{i,j}\right)}{2\alpha+\beta},$$
$$\forall i \in \{1,\ldots,I\},\ j \in \{1,\ldots,J\},\ n \in \{1,\ldots,N\} \quad (2.56)$$

This fast-converging iterative numerical solver removes effectively this signal-dependent noise and overcomes the blurring. Given its 4th-order character, this PDE based filter avoids the blocky effect, and being hyperbolic, it preserves and enhances boundaries and other details. Method comparison results illustrating the effectiveness of the Poisson noise removal technique are shown in Table 2.6 and Fig. 2.9.

2.3 Energy-Based and Nonlinear Diffusion Models for Mixed Denoising

Digital images are often contaminated by mixtures of noises during the capturing process. There are various types of noise mixtures affecting the images, but the most common is the Poisson–Gaussian noise. The 2D Gaussian and quantum noises are often combined into mixed noises that occur when the image sensors have two sources of noise: a signal-dependent one coming from the way the intensity of light is measured and a signal-independent source that is electronic and thermal noise. That mixed noise model is described as $u_0 \cong \mathrm{P}(u) + N\left(0, \sigma^2\right)$, where $u_0 \in L^2(\Omega)$, $\mathrm{P}(u)$ is the image u corrupted by shot noise with distribution $P(n) = \frac{e^{-\mu}\mu^n}{n!}, n \geq 0$ and $N\left(0, \sigma^2\right)$ is the AWGN with zero mean and variance σ.

Various restoration techniques that deal with type of mixed noise have been developed. They include the PURE-LET methods [70], Variance Stabilizing Transform-based algorithms [71], HMM-based approaches [72] and the energy-based models combining total variation schemes for AWGN and quantum filtering [73].

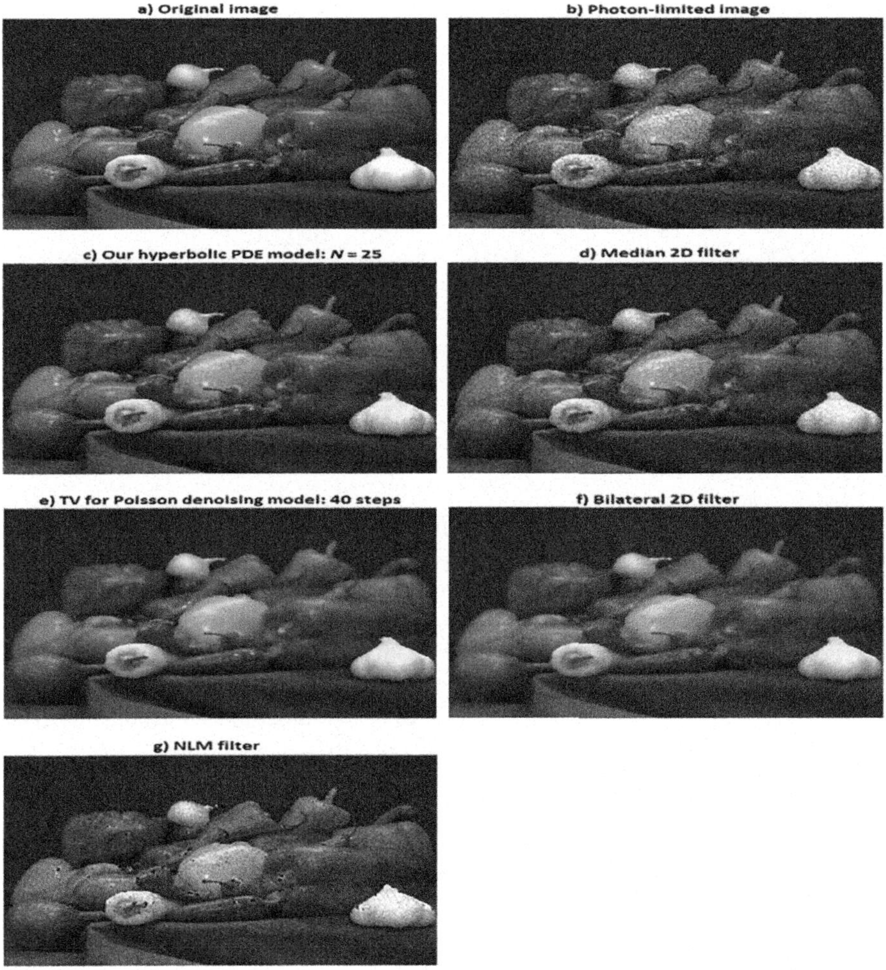

Fig. 2.9 Photon-limited *Peppers* image restored by various filters

Thus, several variational PDE-based mixed noise removal techniques combine TV-ROF Denoising model to some TV schemes adapted for Poisson noise. So, C. T. Pham et al. proposed in 2018 the next energy-based mixed denoising solution:

$$u^* = \arg\min_u \left\{ \int_\Omega |\nabla u| dxdy + \frac{\lambda}{2} \int_\Omega (u - u_0 \log u) dxdy + \beta \int_\Omega (u - u_0 \log u) dxdy \right\} \tag{2.57}$$

where λ, β are the weight parameters of the Gaussian and shot noise and u_0 is the observed image corrupted by the mixed noise [73].

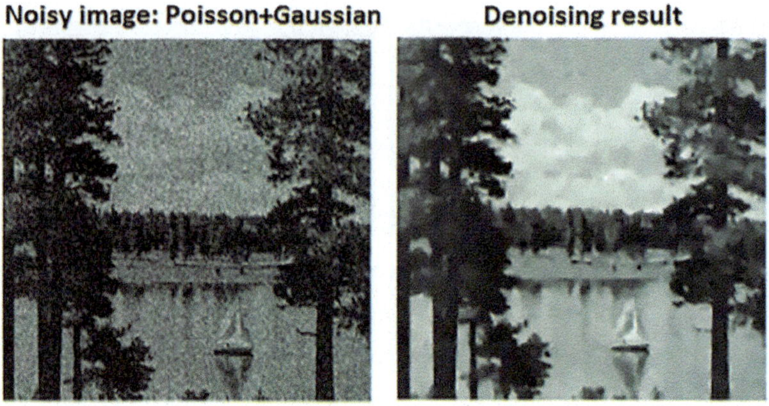

Fig. 2.10 Mixed image denoising using combined TV-based model

A nonlinear PDE model is derived from this variational problem by applying the gradient descent method. A numerical approximation algorithm is created for this model in [73]. This filtering solution removes successfully both the AWGN and the quantum noise. An example of an image deteriorated by Poisson–Gaussian noise that is filtered using this approach is provided in Fig. 2.10.

Another variational mixed noise reduction technique was introduced by Thanh and Dvoenko in 2015 [74]. It combines TV-ROF to a modified TV Denoising. They proposed the mixed denoising problem with constrained conditions:

$$\begin{cases} u^* = \arg\min_{u} \int_{\Omega} |\nabla u| d\Omega \\ \frac{\lambda_1}{2\sigma^2} \int_{\Omega} (u_0 - u)^2 d\Omega + \lambda_2 \int_{\Omega} (u - u_0 \ln u) d\Omega = \kappa \end{cases} \tag{2.58}$$

with the positive regularization parameters $\lambda_1, \lambda_2 > 0, \lambda_1 + \lambda_2 = 1$ and κ is a constant. The optimization problem is solved using the Euler–Lagrange equation and a finite difference-based numerical discretization is applied to the obtained PDE model. This variational filter generates good mixed noise removal results and outperform other filtering approaches, as shown by the denoising example and method comparison results described in Figs. 2.11 and 2.12 [74].

Another energy-based mixed noise removal model uses a spatially adaptive total variation regularization term to enhance the edge preservation ability. The denoising scheme introduced in 2019 by Pham et al. is based on energy minimization:

$$\min_{u,z} \left(\alpha(x) \|z\|_{TV} + \frac{\gamma}{2} \|u - z\|_2^2 + \frac{\lambda_1}{2} \|u - f\|_2^2 + \lambda_2 \langle 1, u - f \log u \rangle \right) \tag{2.59}$$

where z is an auxiliary variable, $\alpha(x) = 1/1 + \frac{|v(x)|}{K}$, $v(x) = |\nabla G_\sigma(x) * f|^2$ and K is a threshold [75]. The Chambolle projection algorithm proposed in 2004 is then applied

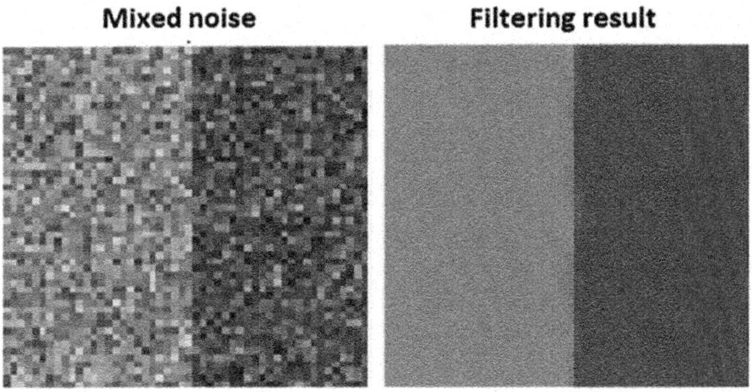

Fig. 2.11 Variational mixed noise reduction example

	Q_{PSNR}	Q_{SSIM}	Q_{MSE}
Noisy	19.5643	0.1036	718.8782
ROF	35.1284	0.9130	19.9635
Median	31.4844	0.7797	46.1996
Wiener	30.1502	0.6018	62.8146
Proposed method with λ_1=0.2, λ_2=0.8, μ = 0.1140, σ = 46.0520	29.1325	0.5933	79.4014
Proposed method with λ_1=λ_2=0.5, μ = 0.1429, σ = 46.0520	37.0462	0.9453	12.8370

Fig. 2.12 Mixed noise method comparison results

to solve numerically this PDE-based scheme [76]. A filtering restoration example based on this approach is displayed in Fig. 2.13.

In Lanza et al. introduced a variational Poisson–Gaussian denoising method using an energy functional combining the TV regularization, Kullback–Leibler divergence term for Poisson noise and the L_2 norm for AWGN [77]. It is based on the next minimization problem:

Fig. 2.13 First row: noisy images; second row: filtered images

$$\hat{f}, \hat{w} = \arg\min_{w,f} \left\{ \frac{1}{2\sigma^2} \|g - w\|_2^2 + D_{KL}(Hf, w) + \lambda TV(f) \right\} \tag{2.60}$$

where the generalized KL divergence of Hf from $w =$ Poisson $(Hf + b1)$ has the form:

$$D_{KL}(Hf, w) = \sum_{i=1}^{N^2} w_i \ln \frac{w_i}{(Hf)_i + b} + (Hf)_i + b - w_i \tag{2.61}$$

This approach provided very good filtering results. An example with some Poisson–Gaussian noise reduction results generated by it is depicted in Fig. 2.14.

A second-order total generalized variation (TGV)-based Poisson denoising model was proposed by Pham et al. in 2021 [78]. It is based on optimization problem

$$u^* = \arg\min_u \left\{ TGV_\alpha^2(u) + \frac{\lambda_1}{2} \int_\Omega (u - f)^2 dxdy + \lambda_2 \int_\Omega (u - f \log u) dxdy \right\} \tag{2.62}$$

where the second-order TGV with weight α is discretized as

$$TGV_\alpha^2(u) = \min_u \alpha_1 \|\nabla u - w\|_1 + \alpha_2 \|\varepsilon(w)\|_1 \tag{2.63}$$

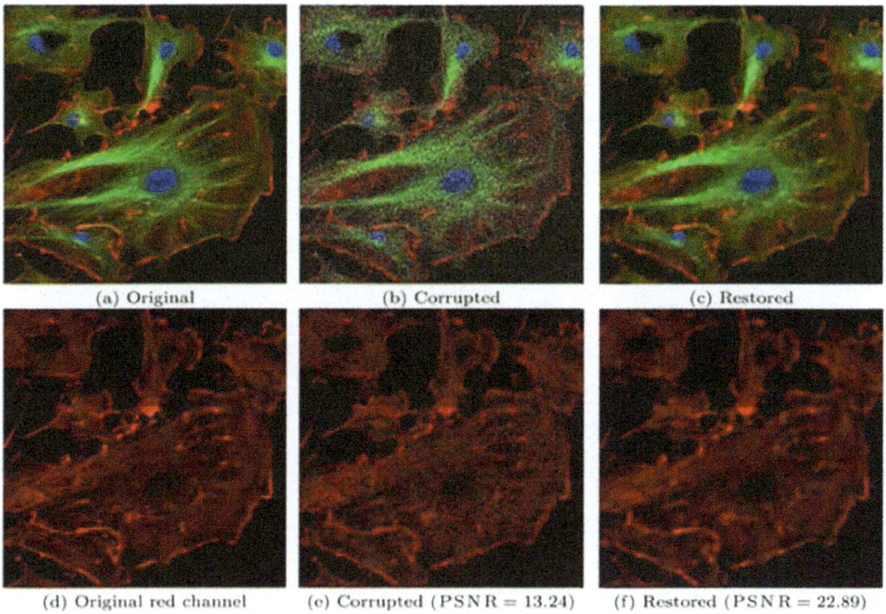

Fig. 2.14 Restoration results provided by the variational Poisson–Gaussian filter

where $w = (w_1, w_2)^{\mathrm{T}}$, $\varepsilon(w) \doteq (1/2)\left(\nabla w + \nabla w^{\mathrm{T}}\right)$. So, minimization problem is discretized as

$$u^* = \arg\min_{u,v}\left(\alpha_1 \|\nabla u - w\|_1 + \alpha_2 \|\varepsilon(w)\|_1 + \frac{\lambda}{2}\|u - f\|_2 + \beta\langle 1, u - f \log u\rangle\right) \tag{2.64}$$

This model was solved by applying an alternating minimization algorithm [78]. It provides a successful quantum-Gaussian filtering and outperforms other mixed denoising schemes, as illustrated by the method comparison results in Fig. 2.15.

We also have several contributions in this PDE-based mixed noise removal field. The first one represents an energy-based restoration technique that combines AWGN and shot noise filtering terms [79]. The proposed variational model is provided by the energy functional minimization $u_{\min} = \arg\min_{u \in BV(\Omega)} E(u)$, where

$$E(u) = \int_{\Omega} \left(\alpha\varphi(\|\nabla u\|) + \lambda(u - u_0 \log u) + \frac{\beta}{2}(u - u_0)^2\right) d\Omega \tag{2.65}$$

and the regularizer function is $\varphi(s) = \int_0^s \tau \nu \sqrt{\frac{\delta}{\gamma\left|\log\left(s^k + \delta\right)\right| + \eta}}\, d\tau$, with $\nu, \lambda, \gamma \in (0, 4]$. One determined the Euler–Lagrange equation related to this variational model and the

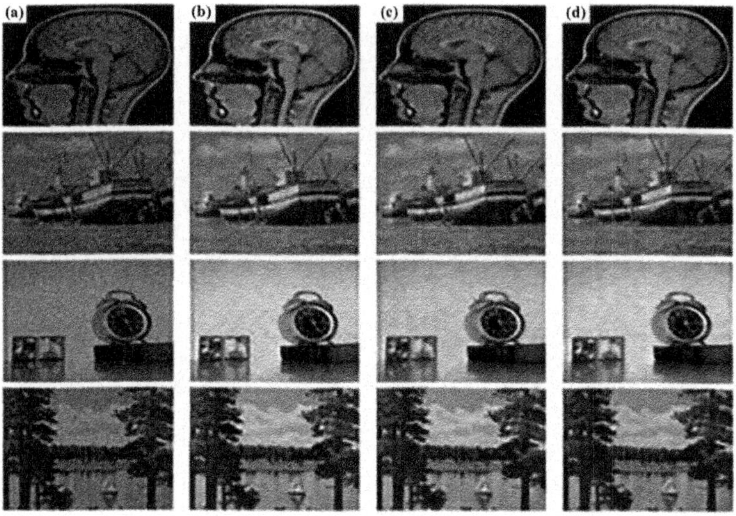

Image	PSNR			
	Noisy	TGV	TVPG	Ours
Board	18.6799	24.0460	24.7064	25.1713
Clock	13.0537	24.3635	24.4234	25.5345
Lake	16.339	22.0954	22.4670	22.8379
Head	16.7107	25.2752	25.6161	26.4411
Average	16.1958	23.9450	24.3032	24.9962

Fig. 2.15 PSNR scores of several TV-based mixed denoising models

gradient descent method was then applied on it. We obtained the nonlinear second-order PDE-based model [79]:

$$\begin{cases} \frac{\partial u}{\partial t} = \alpha \nabla \cdot (\psi(\|\nabla u\|)\nabla u) - (u - u_0)\left(\frac{\lambda}{|u|} + \beta\right), & \forall (x,y) \in \Omega \\ u(x,y,0) = u_0(x,y), \quad \forall (x,y) \in \Omega \subseteq R^2 \\ \frac{\partial u}{\partial \vec{n}}(x,y,t) = 0, \quad \forall (x,y) \in \partial\Omega \end{cases} \tag{2.66}$$

with $\psi(s) = \beta\left(\frac{K}{|\gamma s^k + \zeta|}\right)^{\frac{1}{k+2}}$, $\eta(s) = \varsigma \sqrt[k+2]{\nu s^k + \delta}$, $\beta, \gamma, \zeta \in (0, 5]$, $\nu, \delta \in [1, 5)$ and $\varsigma \in (0, 1]$ [79]. The next explicit iterative numerical approximation scheme was obtained applying the finite difference method on (2.66):

Table 2.7 Average PSNR and SSIM scores obtained by various models

Denoising method	Average PSNR (dB)	Average SSIM
The proposed variational model	31.0581	0.8123
Wiener 2D	25.1425	0.6621
Median filter	24.4826	0.6426
TV-based Poisson denoising	26.7143	0.7432
TV-based mixed noise reduction	28.4268	0.8013

$$\begin{aligned} u_{i,j}^{n+1} &= u_{i,j}^{n}(1-\beta) + u_{i,j}^{0}\left(\frac{\lambda}{u_{i,j}^{n}} + \beta\right) - \lambda + \frac{\alpha}{4} \\ &\cdot \left(\psi\left(\left\|u_{i+1,j}^{n} - u_{i,j}^{n}\right\|\right)\left(u_{i+1,j}^{n} - u_{i,j}^{n}\right) + \psi\left(\left\|u_{i-1,j}^{n} - u_{i,j}^{n}\right\|\right)\left(u_{i-1,j}^{n} - u_{i,j}^{n}\right)\right. \\ &\left. +\psi\left(\left\|u_{i,j+1}^{n} - u_{i,j}^{n}\right\|\right)\left(u_{i,j+1}^{n} - u_{i,j}^{n}\right) + \psi\left(\left\|u_{i,j-1}^{n} - u_{i,j}^{n}\right\|\right)\left(u_{i,j-1}^{n} - u_{i,j}^{n}\right)\right) \end{aligned} \tag{2.67}$$

This PDE-based filters removes successfully both the AWGN and shot noise. It outperforms other mixed noise reduction models, obtaining better performance scores [79]. Some method comparison results are provided in Table 2.7.

Another contribution in this domain is a non-variational PDE-based mixed noise filtering approach [80]. It is based on the next parabolic fourth-order PDE model:

$$\begin{cases} \frac{\partial u}{\partial t} + \lambda\eta(|\nabla u|)\Delta\left(\psi\left(\left\|\nabla^2 u\right\|\right)\Delta u\right) + \alpha(u_0 - u)\left(\frac{1}{|u|+\xi} + 1\right) = 0, & (t,x,y) \in (0,T) \times \Omega \\ u(0,x,y) = u_0(x,y), \quad \forall (x,y) \in \Omega \subseteq R^2 \\ u(t,x,y) = 0, \quad \forall (x,y) \in \partial\Omega \\ \frac{\partial u}{\partial n}(t,x,y) = 0, \quad \forall (x,y) \in \partial\Omega \end{cases} \tag{2.68}$$

where $\psi(s) = \beta\left(\frac{K}{|\gamma s^k + \zeta|}\right)^{\frac{1}{k+2}}$, $\eta(s) = \varsigma \sqrt[k+2]{\nu s^k + \delta}$, $\beta, \gamma, \zeta \in (0, 5]$, $\nu, \delta \in [1, 5)$ and $\varsigma \in (0, 1]$ [80]. The proposed fourth-order PDE-based model was mathematically treated, its validity being demonstrated rigorously in [80]. This PDE is well-posed, since a unique weak solution exists for it. The next explicit and stable numerical approximation scheme algorithm was proposed as PDE model's solver [80]:

$$
\begin{aligned}
u_{i,j}^{n+1} = {} & u_{i,j}^{n}\left(\frac{1}{\left|u_{i,j}^{n}\right| + \xi}\right) - \lambda\eta\left(\left\|\nabla u_{i,j}^{n}\right\|\right)\left(\varphi_{i+1,j}^{n} - \varphi_{i-1,j}^{n} - \varphi_{i,j+1}^{n} - \varphi_{i,j-1}^{n} + 4\varphi_{i,j}^{n}\right) \\
& + u_{i,j}^{0}\alpha\left(\frac{1}{\left|u_{i,j}^{n}\right| + \xi} + 1\right)
\end{aligned}
\tag{2.69}
$$

The method comparison results illustrating the effectiveness of the proposed technique are described in Fig. 2.16 and Table 2.8 [80].

Fig. 2.16 Results achieved by several mixed denoising methods

Table 2.8 Average PSNRs for several denoising techniques

Filtering method	Average PSNR (dB)
The proposed PDE-based model	27.4531
2D median filter	23.3912
Wiener filter	24.7425
TV-ROF denoising	25.3147
2nd order PDE-based mixed noise removal	26.4369

2.4 Multilayer Image Denoising Using Vector-Valued PDE-Based Models

The digital images may have a single channel, like the grayscale images, or multiple channels. There are various types of multi-channel, or multi-layer, images, including the digital images in various color spaces (RGB, HSV, CIELAB, CMYK and others), the spectral images of various types, the multimodal and multi-parametric images. They are characterized by at least 3 layers, but may have even hundreds of channels.

Since the capturing instruments involve various physical measurements, these images could be contaminated by multiple noises during the acquisition process, including AWGN, quantum noise, impulse noise and noise mixtures. Their restoration still represents a challenging image processing task. Since their channels are highly correlated, filtering each one separately would not provide a proper result. An effective restoration has to address properly the inter-channel correlation issue.

The filtering of the most common multi-channel images, which are those in the RGB format, will be discussed later in this book, in the multi-scale analysis context. Here we focus on the most important category of multilayer images, which is that of *chemical*, or spectral, *images* [81]. Spectral imaging process collects the image data at different wavelengths across the electromagnetic spectrum. It can use the infrared, visible spectrum, X-rays, ultra-violet or various combination of the above. The wavelengths are identified by using devices that are sensitive to particular wavelengths, like the light from frequencies beyond visible light range. Spectral sensors acquire the infrared and ultra-violet lights that are invisible to the human eye. Thus, the spectral imagery enable the observation of not so apparent things. The main spectral imaging types are the multispectral (MSI) and the hyperspectral imaging (HSI). Other spectral imaging types include the ultraspectral imaging that uses interferometer sensors characterized by a very fine spectral resolution.

Since multispectral imaging collects the spectral information from a limited number of frequencies, or wavelengths, while recording the image at the same time, a multispectral image represents a collection of few mono-chrome layers of the same scene, each one captured at a certain frequency band. The number of those MSI bands is usually in the [3, 15] interval and the wavelength range is uncontinuous, since the MSI contain information in discrete and specific wavelengths [82].

Hyperspectral imaging collects the spectral information across a wide range of wavelengths while simultaneously capturing the spatial information of the image, and provides a complete spectrum for each pixel. So, HSI represent chemical images acquired in several hundreds of continuous and contiguous spectral bands [83]. The main application areas of the spectral images include the medical and astronomical imaging, remote sensing, fluorescence microscopy, document and artwork investigation, military target tracking, maritime surveillance, earth exploring, weather prediction and thermal signature detection [84].

Numerous spectral image filtering solutions have been developed for various types of noise. The AWGN removal techniques include methods based on transforming images into another space by means like sparse representation [85], the 2D filters adapted for volumetric images, like the 3D NLM filter [86], 3D Gaussian filter, 3D bilateral filter [87], the tensor-based techniques [88], the adaptive Wiener filters [89], the block-matching and 3D filtering (BM3D) models [90], the PDE-based models and deep learning solutions [91]. The photon-limited vector-valued image denoising approaches include the 3D Median filter, 3D NLM Poisson filter [92], 3D Gaussian filter with VST [93], tensor-based filters [88], spectral unmixing-based models [94] and energy-based schemes [95]. The mixed noise removal methods for spectral images include spatial-spectral structure similarity-based techniques [96], tensor-based schemes [97], some extensions of the 2D filters [98] and the generalized Anscombe transformation [93].

Many extensions of the nonlinear PDEs have been used successfully in this image restoration field. So, some vector-valued PDE-based filtering schemes, which introduce a coupling between image channels, have been proposed for the multi-valued images [99–101]. Thus, a nonlinear anisotropic diffusion-based hyperspectral image filtering technique is described in [99]. First, the data is projected in a transformed space with a Maximum Noise Fraction (MNF) transform, allowing the new components to be sorted in order of their signal to noise (SNR) ratio. The filtering is then applied on each component in the MNF space with the use of the SNR estimation to control the feature scale. A vectorial anisotropic diffusion algorithm, which represents an extended version of the anisotropic diffusion model of Perona and Malik based on a measure of spectral similarity S between two vectors is applied to the transformed image, I_{MNF}, as follows:

$$I_{MNF_{i,j}}^{t+1} = I_{MNF_{i,j}}^{t} + \frac{1}{4}\sum_{d=1}^{4} c_{d_{i,j}}^{t} \nabla_d I_{MNF_{i,j}}^{t} \tag{2.70}$$

where $c_{d_{i,j}}^{t} = g\left(S_d I_{i,j}^{t}\right)$, with g being that used in (2.2) [99].

A vector-valued diffusion-based approach is proposed by Whitaker in [100]. It defines the vector-valued image as $\vec{F} : I \times T \mapsto D$, where D is the feature space, and the considered vector-valued diffusion equation is

$$\nabla \cdot g\left(D_s \vec{F}\right) \nabla \vec{F} = \frac{\partial \vec{F}}{\partial t}, \tag{2.71}$$

where the vectorial dissimilarity operator D_s is the root sum of squares of the Jacobian matrix of $\vec{F}$ [100].

This vector-valued nonlinear diffusion-based filtering provided successful results when applied to 3-D double-echo MR image data [100]. A MRI restoration example is described in Fig. 2.17.

Variational vector-valued image regularization solutions have been also proposed. Such a multi-valued image filtering is based on the functional minimization:

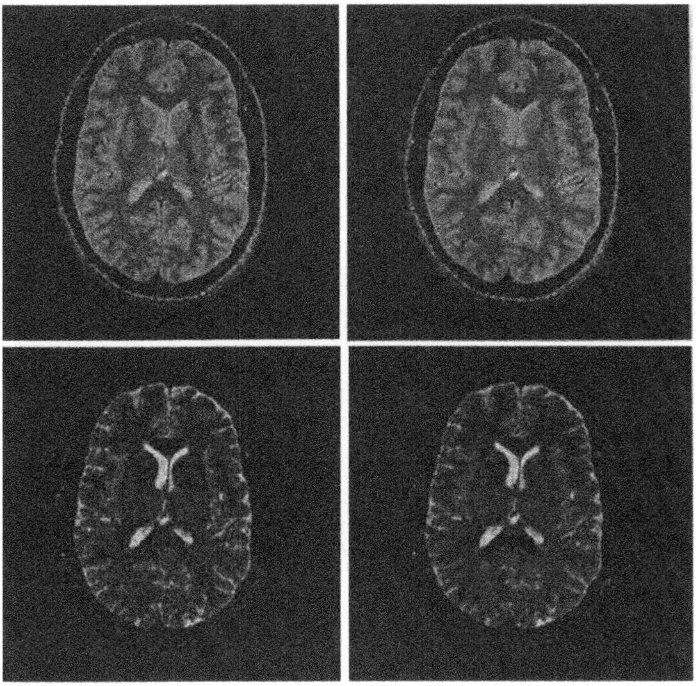

Fig. 2.17 Example of filtering 3-D double-echo MR data

$$\min_{I:\Omega\to R^n} E(I) = \int_{\Omega} \phi(\mathcal{N}(I))d\Omega, \quad (2.72)$$

where ϕ is an increasing function and norm $\mathcal{N}(I)$ is the trace of the structure tensor:

$$G = \sum_i \nabla I_i \nabla I_i^T, \quad (2.73)$$

where I_i are components of I [101]. The minimization in (2.72) is performed using a gradient descent given by the Euler–Lagrange equations of $E(I)$. These equations lead to a system of divergence-based expressions

$$\frac{\partial I_i}{\partial t} = div(D\nabla I_i), \quad i = 1, 2, \ldots \quad (2.74)$$

where the $[2 \times 2]$ diffusion tensor D is a symmetric and positive matrix designed from the spectral elements of G [101]. Other similar tensor-based anisotropic diffusion filters for vector-valued images are described in [102].

We have also developed vector-valued PDE-based restoration frameworks for various types of multi-valued images and noises. A vector-valued diffusion-based multi-spectral image denoising approach that filters successfully the AWGN is introduced in [103]. It is based on the next vector-valued hyperbolic PDE model:

$$\begin{cases} \alpha \frac{\partial^2 \vec{u}}{\partial t^2} + \beta \frac{\partial \vec{u}}{\partial t} - \lambda \psi\left(\frac{|\Delta \vec{u}|+|\nabla \vec{u}|}{3}\right) \nabla \cdot (\varphi(\|\nabla \vec{u}_\sigma\|)\nabla \vec{u}) + \gamma(\vec{u} - \vec{u}_0) = 0, \\ \vec{u}(x,y,0) = \vec{u}_0(x,y), \quad \forall (x,y) \in \Omega \\ \vec{u}_t(x,y,0) = \vec{u}_1(x,y), \quad \forall (x,y) \in \Omega \\ \vec{u}(x,y,t) = 0, \quad \forall (x,y) \in \partial\Omega, \quad t \in (0,T) \end{cases} \tag{2.75}$$

where $\alpha, \beta \in [1, 2)$, $\lambda \in (0, 1]$, $\gamma \in (0, 0.5)$, the vector-valued function representing the spectral data with K layers is $\vec{u} : \Omega \times (0, T) \to R^K$, where

$$\vec{u}(x,y) = \left[u^1(x,y), \ldots, u^K(x,y)\right], \quad \forall (x,y) \in \Omega, \tag{2.76}$$

the conductance function $\varphi(s) = \xi\left(\frac{\eta}{|\delta \log 10(s+\eta)^2 + \varepsilon|}\right)^{\frac{1}{3}}$, $\delta \in (0, 1]$, $\xi \geq 1$, $\eta, \varepsilon \geq 5$ and $\psi(s) = \varsigma \sqrt[\tau+1]{\zeta s^\tau + \nu}$, with $\zeta, \nu \in [1, 4)$, $\varsigma \in (0.5, 1)$, $\tau \in \{1, \ldots, 4\}$ [103].

The conductance $\varphi(\|\nabla \vec{u}_\sigma\|)$ and term $\psi\left(\frac{|\Delta \vec{u}|+|\nabla \vec{u}|}{3}\right)$ couple the single-valued PDEs generated by (2.75) in order to deal properly with the inter-channel correlation issue. Since $\nabla \vec{u}$ is the generalized Jacobian of $\vec{u}$, the divergence term becomes:

$$\nabla \cdot (\varphi(\|\nabla \vec{u}_\sigma\|)\nabla \vec{u}) = \left(\frac{\partial}{\partial x}, \frac{\partial}{\partial y}\right) \cdot \varphi(\|\nabla \vec{u}_\sigma\|) \begin{pmatrix} \frac{\partial u_1}{\partial x} & \cdots & \frac{\partial u_K}{\partial x} \\ \frac{\partial u_1}{\partial y} & \cdots & \frac{\partial u_K}{\partial y} \end{pmatrix}, \tag{2.77}$$

Since $\|\nabla \vec{u}_\sigma\| = \sqrt{\sum_{k=1}^K \left\|\nabla u_\sigma^k\right\|^2}$, $|\nabla \vec{u}| = \sqrt{\sum_{k=1}^K \left\|\nabla u^k\right\|^2}$ and $|\Delta \vec{u}| = \sum_{k=1}^K \left\|\Delta u^k\right\|$, the vector-valued PDE model leads to the next system of single-valued hyperbolic PDEs:

$$\begin{cases} \alpha \frac{\partial^2 u^1}{\partial t^2} + \beta \frac{\partial u^1}{\partial t} - \lambda \psi\left(\frac{\sum_{k=1}^K \left\|\Delta u^k\right\| + \sqrt{\sum_{k=1}^K \left\|\nabla u^k\right\|^2}}{3}\right) div\left(\varphi\left(\sqrt{\sum_{k=1}^K \left\|\nabla u_\sigma^k\right\|^2}\right)\nabla u^1\right) + \gamma\left(u^1 - u_0^1\right) = 0, \\ \vdots \\ \alpha \frac{\partial^2 u^K}{\partial t^2} + \beta \frac{\partial u^K}{\partial t} - \lambda \psi\left(\frac{\sum_{k=1}^K \left\|\Delta u^k\right\| + \sqrt{\sum_{k=1}^K \left\|\nabla u^k\right\|^2}}{3}\right) div\left(\varphi\left(\sqrt{\sum_{k=1}^K \left\|\nabla u_\sigma^k\right\|^2}\right)\nabla u^K\right) + \gamma\left(u^K - u_0^K\right) = 0 \end{cases} \tag{2.78}$$

The second-order hyperbolic equations evolve simultaneously but not independently, sharing channel correlation terms $\varphi\left(\sqrt{\sum_{k=1}^K \left\|\nabla u_\sigma^k\right\|^2}\right)$ and $\psi\left(\frac{\sum_{k=1}^K \left\|\Delta u^k\right\| + \sqrt{\sum_{k=1}^K \left\|\nabla u^k\right\|^2}}{3}\right)$. These correlations model properly the MSI inter-channel relations. So, the multi-edges of the spectral image are preserved well both along and across its layers [103].

Each single-valued PDE is non-variational and well-posed, admitting a unique weak solution. The solution of the vector-valued PDE model is determined by solving numerically the system (2.78). So, the PDE related to u^κ, $\forall\kappa \in \{1, \ldots, K\}$, is:

$$\alpha\frac{\partial^2 u^\kappa}{\partial t^2} + \beta\frac{\partial u^\kappa}{\partial t} + \gamma\left(u^\kappa - u_0^\kappa\right) = \lambda\psi\left(\frac{\sum_{k=1}^K \left|\Delta u^k\right| + \sqrt{\sum_{k=1}^K \left|\nabla u^k\right|^2}}{3}\right)\nabla \cdot \left(\varphi\left(\sqrt{\sum_{k=1}^K \left|\nabla u_\sigma^k\right|^2}\right)\nabla u^\kappa\right) \tag{2.79}$$

The following discretization of this equation is determined in [103] for $n \in \{0, \ldots, N\}, i \in \{1, \ldots, I\}, j \in \{1, \ldots, J\}$:

$$\left(u^\kappa\right)_{i,j}^{n+1}\left(\alpha + \frac{\beta}{2}\right) = \left(u^\kappa\right)_{i,j}^{n-1}\left(\frac{\beta}{2} - \alpha\right) + \left(u^\kappa\right)_{i,j}^{n}(2\alpha - \gamma)$$
$$+\gamma\left(u^\kappa\right)_{i,j}^{0} + \lambda\psi\left(\frac{1}{3}\left(\begin{array}{c}\sum_{k=1}^K\left(\left(\left(u^k\right)_{i+1,j}^n + \left(u^k\right)_{i-1,j}^n + \left(u^k\right)_{i,j+1}^n + \left(u^k\right)_{i,j-1}^n - 4\left(u^k\right)_{i,j}\right)\right) \\ +\frac{1}{2}\sqrt{\sum_{k=1}^K\left(\left(u^k\right)_{i+1,j}^n - \left(u^k\right)_{i-1,j}^n\right)^2 + \left(\left(u^k\right)_{i,j+1}^n - \left(u^k\right)_{i,j-1}^n\right)^2}\end{array}\right)\right)\cdot$$
$$\rho\sum_{r\in N(p)}\varphi\left(\frac{1}{2}\sqrt{\sum_{k=1}^K\left(\left(u_\sigma^k\right)_{i+1,j}^n - \left(u_\sigma^k\right)_{i-1,j}^n\right)^2 + \left(\left(u_\sigma^k\right)_{i,j+1}^n - \left(u_\sigma^k\right)_{i,j-1}^n\right)^2}\right)\left(\left(u^\kappa\right)_r^n - \left(u^\kappa\right)_p^n\right) \tag{2.80}$$

The numerical solution of the vector-valued nonlinear PDE model (2.75) is computed applying (2.80) iteratively for each n from 0 to N and κ from 1 to K [103]. This restoration approach removes successfully the additive Gaussian noise, overcomes unintended blurring and staircase effects and preserves well, and even sharpens, the essential details among the different wavelength bands. A MSI restoration example is displayed in Fig. 2.18 and Table 2.9 contains method comparison results.

A photon-limited multi-channel image restoration technique was proposed in [104]. It uses the next well-posed vector-valued fourth-order reaction–diffusion model:

$$\begin{cases}\frac{\partial\vec{u}}{\partial t} + \alpha\delta\left(\frac{\|\nabla^2\vec{u}\|}{\|\nabla\vec{u}\|}\right)\nabla^2\left(\psi\left(\|\nabla\vec{u}_\sigma\|\right)\Delta\vec{u}\right) - \beta\left(\frac{\vec{u}-\vec{u}_0}{|\vec{u}|+\lambda}\right) = 0, \\ \vec{u}(x, y, 0) = \vec{u}_0(x, y), \quad \forall(x,y) \in \Omega \\ \frac{\partial\vec{u}}{\partial\vec{n}}(x, y, t) = 0, \quad \forall(x,y) \in \partial\Omega\end{cases} \tag{2.81}$$

where $\alpha \in [1, 2)$, $\lambda, \beta \in (0, 1]$, $\vec{u}_0$ is the observation corrupted by Poisson noise, $\vec{u}(x, y) = \left[u^1(x, y), \ldots, u^C(x, y)\right]$, $\psi(s) = \eta\left(\frac{\gamma}{|\varphi(s+\gamma)^\tau + \xi|}\right)^{\frac{1}{\tau+1}}$, $\varphi \in (0, 1]$, $\eta \geq 1$, $\gamma, \xi \geq 4$, $\tau \in \{1, \ldots, 4\}$, $\delta(s) = \varepsilon\sqrt[r+1]{\nu s^r + \varsigma}$, with $\varsigma, \nu \in [1, 4)$, $\varepsilon \in (0, 1)$ and $r \in [1, 3]$. It leads to the next nonlinear PDE system:

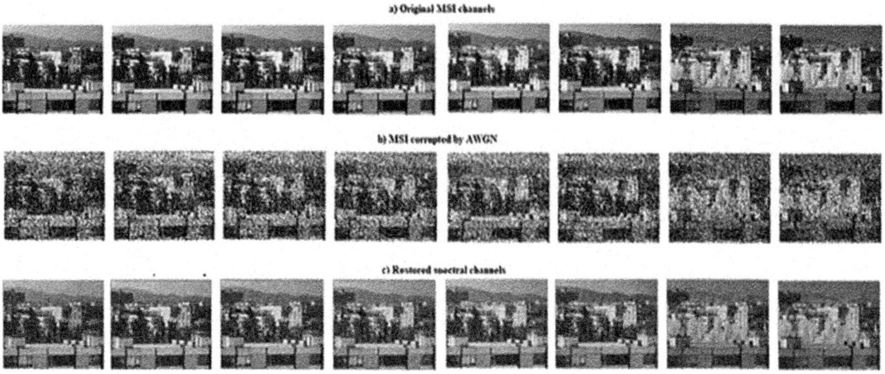

Fig. 2.18 MSI restoration example

Table 2.9 Average PSNRs of several AWGN removal methods

Noise removal technique	PSNR score (dB)
Vector-valued hyperbolic PDE-based filter	27.2532
Adaptive Wiener filter	26.8443
3D Gaussian filter	25.4778
3D non-local mean (NLM) filter	27.2348
3D bilateral filter	26.5738
TV-ROF denoising	25.4673

$$\begin{cases} \frac{\partial u^1}{\partial t} = \beta\left(\frac{u^1 - u_0^1}{|u^1| + \lambda}\right) - \alpha\delta\left(\frac{\sum_{k=1}^{C}|\Delta u^k|}{\sqrt{\sum_{k=1}^{C}|\nabla u^k|^2}}\right)\Delta\left(\psi\left(\sqrt{\sum_{k=1}^{C}\|\nabla u_\sigma^k\|^2}\right)\Delta u^1\right), \\ \vdots \\ \frac{\partial u^C}{\partial t} = \beta\left(\frac{u^C - u_0^C}{|u^C| + \lambda}\right) - \alpha\delta\left(\frac{\sum_{k=1}^{C}|\Delta u^k|}{\sqrt{\sum_{k=1}^{C}|\nabla u^k|^2}}\right)\Delta\left(\psi\left(\sqrt{\sum_{k=1}^{C}\|\nabla u_\sigma^k\|^2}\right)\Delta u^C\right) \end{cases} \tag{2.82}$$

that is solved numerically by applying the next iterative fast-converging finite difference-based discretization scheme for $\forall n \in \{0, \ldots, N\}$ and each k from 1 to C [104]:

$$\left(u^k\right)_{i,j}^{n+1} = \left(u^k\right)_{i,j}^{n}\left(\frac{\beta}{\left|\left(u^k\right)_{i,j}^{n}\right| + \lambda} + 1\right) - \left(u^k\right)_{i,j}^{0}\left(\frac{\beta}{\left|\left(u^k\right)_{i,j}^{n}\right| + \lambda}\right)$$
$$- \delta\left(\frac{\sum_{k=1}^{C}\left(u^k\right)_{i+1,j}^{n} + \left(u^k\right)_{i-1,j}^{n} + \left(u^k\right)_{i,j+1}^{n} + \left(u^k\right)_{i,j-1}^{n} - 4\left(u^k\right)_{i,j}^{n}}{\sqrt{\sum_{k=1}^{C}\left\|\nabla\left(u^k\right)_{i,j}^{n}\right\|^2}}\right)$$
$$\left(\zeta^{i+1,j} + \zeta^{i-1,j} + \zeta^{i,j+1} + \zeta^{i,j-1} - 4\zeta^{i,j}\right) \tag{2.83}$$

Our quantum denoising simulations were performed on 2 multi-valued image databases. The testing dataset contained 200 multi-channel images, representing 100 8-channel multispectral images from the UGR Spectral Saliency Database and 100 RGB images from the IPL Calibrated Color Image Database [104], affected by high levels of Poisson noise. The described filter removed successfully the signal-dependent noise from these images, avoiding the side-effects and preserving the essential features. Also, it addressed properly the high inter-channel correlation, preserving the meaningful multi-boundaries among the different layers. It outperforms other multi-channel photon-limited image filters, as illustrated by the method comparison results in Table 2.10 [104].

Other multi-layer mage restoration approaches developed by us deal successfully with the mixed Poisson–Gaussian noise [105, 106]. Thus, a noise mixture removal for multispectral images with M layers was introduced in [105]. It is based on the following second-order vector-valued diffusion model:

$$\begin{cases} \frac{\partial \vec{u}}{\partial t} - \alpha\varphi\left(\left|\nabla^2\vec{u}\right|\right)\nabla\cdot\left(\psi\left(\left\|\nabla\vec{u}_\sigma\right\|\right)\nabla\vec{u}\right) + \left(\vec{u} - \vec{u}_0\right)\left(\frac{\beta}{|\vec{u}|+\varepsilon} + \lambda\right) = 0, \\ \vec{u}(x,y,0) = \vec{u}_0(x,y), \quad \forall (x,y) \in \Omega \\ \frac{\partial \vec{u}}{\partial \vec{n}}(x,y,t) = 0, \quad \forall (x,y) \in \partial\Omega, \quad t \in (0,T) \end{cases} \tag{2.84}$$

where $\alpha \in [1,2), \beta, \lambda \in (0,1], \varepsilon \in (0,0.5), \vec{u} : \Omega \times (0,T) \to R^M$, $\vec{u}_\sigma = [u_k * G_\sigma]_{k=\overline{1,M}}$, $\psi(s) = \gamma\sqrt[3]{\frac{K}{\left|\nu \ln(s+K)^2 + \xi\right|}}$, $\nu \in (0,1], \gamma \geq 1, \xi \geq 5, K \geq 7$ and

Table 2.10 Photon-limited multi-layer image restoration method comparison

Poisson noise removal technique	PSNR value (dB)
Vector-valued reaction diffusion-based filter	28.6872
3D median filter	26.5748
3D Gaussian filter + VST	25.3966
3D NLM Poisson filter	27.9354
3D bilateral filter	26.9783
TV denoising for Poisson noise	25.7558

$\varphi(s) = \left(\frac{\delta s^r + \zeta}{\eta}\right)^{\frac{1}{r+1}}$, with $\delta, \zeta \in [1, 5)$, $\eta \in (0, 0.5)$ and $r \in (0, 1)$. The conductance term $\psi(\|\nabla \vec{u}_\sigma\|)$ and $\varphi(|\nabla^2 \vec{u}|) = \varphi(|\Delta \vec{u}|)$ couple of the single-diffusion processes resulted from (2.84). Like in the previous cases, a nonlinear PDE-based system is obtained from this model. That system is solved by applying this explicit iterative fast-converging numerical approximation scheme for each n from 0 to N and for each m from 1 to M [105]:

$$
\begin{aligned}
\left(u^m\right)_{i,j}^{n+1} &= \left(u^m\right)_{i,j}^{n}\left(\lambda - \frac{\beta}{\left|\left(u^m\right)_{i,j}^{n}\right| + \varepsilon} + 1\right) + \left(u^m\right)_{i,j}^{0}\left(\frac{\beta}{\left|\left(u^m\right)_{i,j}^{n}\right| + \varepsilon} + \lambda\right) \\
&+ \alpha\varphi\left(\sum_{k=1}^{M}\left(u^k\right)_{i+1,j}^{n} + \left(u^k\right)_{i-1,j}^{n} + \left(u^k\right)_{i,j+1}^{n} + \left(u^k\right)_{i,j-1}^{n} - 4\left(u^k\right)_{i,j}^{n}\right) \\
&\zeta\sum_{q\in N_p}\psi\left(\frac{1}{2}\sqrt{\sum_{k=1}^{M}\left(\left(u_\sigma^k\right)_{i+1,j}^{n} - \left(u_\sigma^k\right)_{i-1,j}^{n}\right)^2 + \left(\left(u_\sigma^k\right)_{i,j+1}^{n} - \left(u_\sigma^k\right)_{i,j-1}^{n}\right)^2}\right) \\
&\left(\left(u^m\right)_{q}^{n} - \left(u^m\right)_{p}^{n}\right)
\end{aligned} \tag{2.85}
$$

This approach models efficiently inter-channel relations and capture the multi-edges, providing an essential detail-preserving MSI filtering. It deals properly with both signal-independent and -dependent noise components from the MSI channels, removing successfully both the AWGN and shot noise. Our noise mixture removal scheme outperforms the 2D restoration methods that filter each layer separately and also some volumetric 3D filters [105]. Figure 2.19 displays an example with its results.

A mixed noise removal framework for hyperspectral images was introduced by us in [106]. It uses the vector-valued second-order nonlinear diffusion-based model:

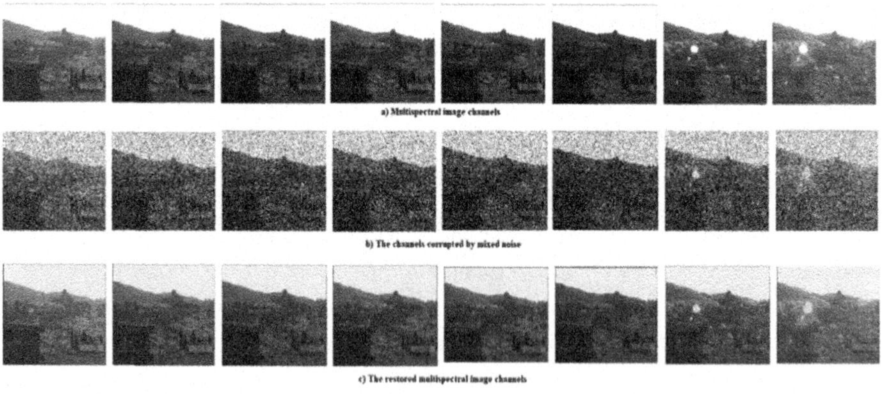

Fig. 2.19 A MSI filtering example

$$\begin{cases} \frac{\partial \vec{u}}{\partial t} - \lambda \nabla \cdot (\delta(\|\nabla \vec{u}\|) \nabla \vec{u}_\sigma) \varphi\left(\frac{|\nabla \vec{u}| + |\nabla^2 \vec{u}|}{\xi}\right) + \left(\frac{\psi}{\|\vec{u}\| + \alpha} + \zeta\right)(\vec{u} - \vec{u}_0) = 0, \\ \vec{u}(x, y, 0) = \vec{u}_0(x, y), \quad \forall (x,y) \in \Omega; \quad \vec{u}(x, y, t) = 0, \quad \forall (x,y) \in \partial\Omega, \quad t \in (0,T) \end{cases} \tag{2.86}$$

where $\lambda \in [1, 3), \psi, \zeta \in (0, 1], \lambda \in (0, 0.6), \xi \in \{1, \ldots, 4\}$, $\vec{u}_\sigma = \left[u^1 * G_\sigma, \ldots, u^H * G_\sigma\right]$, the conductance function $\delta(s) = \eta \sqrt[\xi+1]{\frac{\gamma}{|\nu \log_{10}(s+\gamma)^\xi + \varepsilon|}}$ and $\varphi(s) = \left(\frac{\varsigma s^\xi + r}{\lambda}\right)^{\frac{1}{\xi+1}}$, where $\eta \geq 1, \nu \in (0, 1], \gamma, \varepsilon \geq 3, r, \varsigma \in [1, 4)$ [106]. The nonlinear multi-valued PDE model (2.86) leads to the following reaction–diffusion system:

$$\begin{cases} \frac{\partial u^1}{\partial t} - \lambda \varphi\left(\frac{\sqrt{\sum_{k=1}^H \left\|\nabla u^k\right\|^2} + \sum_{k=1}^H \left\|\Delta u^k\right\|}{\xi}\right) div\left(\delta\left(\sqrt{\sum_{k=1}^H \|\nabla u^k\|^2}\right) \nabla u^1_\sigma\right) + \left(\frac{\psi}{\left\|u^1\right\| + \alpha} + \zeta\right)\left(u^1 - u^1_0\right) = 0, \\ \vdots \\ \frac{\partial u^H}{\partial t} - \lambda \varphi\left(\frac{\sqrt{\sum_{k=1}^H \left\|\nabla u^k\right\|^2} + \sum_{k=1}^H \left\|\Delta u^k\right\|}{\xi}\right) div\left(\delta\left(\sqrt{\sum_{k=1}^H \|\nabla u^k\|^2}\right) \nabla u^H_\sigma\right) + \left(\frac{\psi}{\left\|u^H\right\| + \alpha} + \zeta\right)\left(u^H - u^H_0\right) = 0 \end{cases} \tag{2.87}$$

Its coupled single-valued reaction–diffusion equations share 2 correlation terms that model the HSI inter-layer relations: $\varphi\left(\left(\sqrt{\sum_{k=1}^H \left\|\nabla u^k\right\|^2} + \sum_{k=1}^H \left\|\Delta u^k\right\|\right)/\xi\right)$ and $\delta\left(\sqrt{\sum_{k=1}^H \left\|\nabla u^k\right\|^2}\right)$.

An iterative finite difference-based numerical approximation algorithm that solves this system is proposed in [106]. Then, it was tested successfully on many HSI databases [106], including *Real-World Hyper-spectral Images Database*, *Granada hyperspectral hazy database* (*GRANHHADA*), *UGR Hyperspectral Image Data-base* and *HIDSAG*. Our HSI restoration approach filters effectively the noise mixtures from hyperspectral data and overcomes side-effects like blurring and staircasing, preserving essential details. It also deals successfully with inter-channel correlation issue, preserving well meaningful features, such as the multi-edges, across the hyperspectral layers. See some comparison results in Table 2.11 and Fig. 2.20.

Table 2.11 HSI filtering method comparison

HSI restoration solution	Average PSNR (dB)
Multi-valued reaction diffusion-based filter	27.8476
3D non-local mean (NLM) filter	27.4533
TV-based mixed noise reduction model	26.5627
TV denoising + VST scheme	25.6618
3D Gaussian filter + VST	25.3754
3D median filter	24.5209

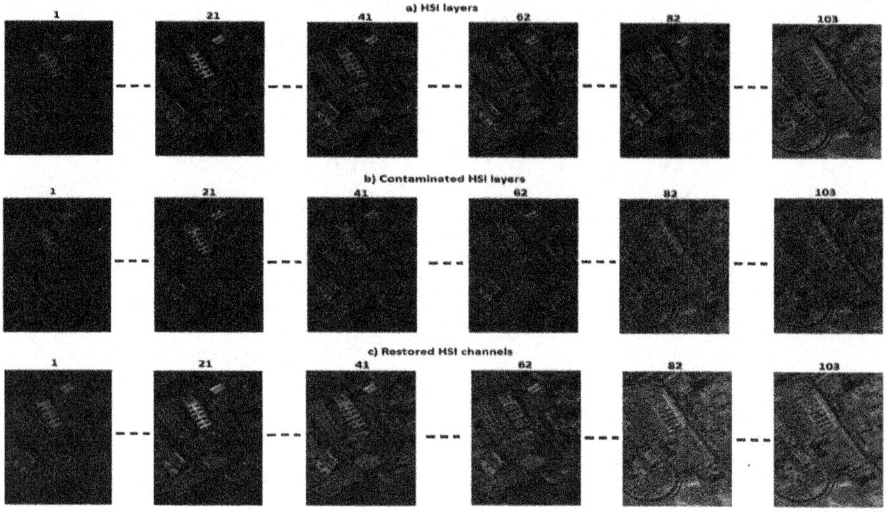

Fig. 2.20 HSI data restoration example

Another mixed noise reduction approach was proposed by us for the multi-modal images [107]. So, the introduced multimodal *magnetic resonance image* (MRI) restoration technique uses the next vector-valued anisotropic diffusion-based model:

$$\begin{cases} \frac{\partial \vec{u}}{\partial t} - \xi \psi\left(\frac{|\nabla \vec{u}| + |\Delta \vec{u}|}{2}\right) \nabla \cdot (\varphi(\|\nabla \vec{u}_\sigma\|) \nabla \vec{u}) + \left(\frac{\alpha}{|\vec{u}| + \lambda} + \beta\right)(\vec{u} - \vec{u}_0) = 0, \\ \vec{u}(x, y, 0) = \vec{u}_0(x, y), \quad \forall (x, y) \in \Omega \\ \frac{\partial \vec{u}}{\partial \vec{n}}(x, y, t) = 0, \quad \forall (x, y) \in \partial\Omega, \quad t \in (0, T) \end{cases} \tag{2.88}$$

where $\vec{u} : \Omega \times (0, T) \to R^L$ is a multimodal 3D MRI, $\xi \in [1, 2), \alpha, \beta \in (0, 1], \lambda \in (0, 0.5)$, the monotonic decreasing conductance function $\varphi(s) = \delta\left(\frac{\nu}{|\varepsilon \ln(s+\nu)^r + \eta|}\right)^{\frac{1}{r+1}}$ and $\psi(s) = \gamma\sqrt[3]{\varsigma s^2 + \zeta}$, with $\varepsilon \in (0, 1], \delta \geq 1, \nu, \eta \geq 4, r \in \{1, \ldots, 4\}, \zeta, \varsigma \in [1, 3), \gamma \in (0.5, 1)$.

This vector-valued PDE model leads to a system with equations of the form:

$$\frac{\partial u^l}{\partial t} + \left(\frac{\alpha}{|u^l| + \lambda} + \beta\right)(u^l - u_0^l) = \xi \psi\left(\frac{\sqrt{\sum_{k=1}^L \|\nabla u^k\|^2} + \sum_{k=1}^L \|\Delta u^k\|}{2}\right) \nabla \cdot \left(\varphi\left(\sqrt{\sum_{k=1}^L \|\nabla u_\sigma^k\|^2}\right) \nabla u^l\right), \quad l \in \{1, \ldots, L\} \tag{2.89}$$

which are then solved numerically by applying the finite difference method [107].

The vector-valued anisotropic diffusion-based MRI denoising method was tested successfully on a voluminous medical dataset [107]. It filters successfully both the AWGN and shot noise from those 3D medical images and overcomes the side-effects. It also addresses properly the inter-layer correlation issue, preserving essential features, like the multi-edges, among and across the MRI channels. A MRI filtering example is provided in Fig. 2.21 and Table 2.12 contains method comparison results.

2.5 Nonlinear 3D Anisotropic Diffusion-Based Video Filtering Solutions

Video noise, representing a random variation in the image's brightness, constitutes an unintended effect of the video acquisition process, being typically a result of the digital gain [108]. While the most common type of video noise is the signal-independent AWGN, there are also other noises generated by the video capturing devices, including signal-dependent quantum noise, multiplicative noise, stripe noise, impulse noise and various mixed noises.

Removing this noise from the movie frames represents an essential pre-processing operation that facilitates many video analysis tasks. But video filtering may represent also a difficult task, since the consecutive frames are highly correlated. The spatial video denoising techniques, that filter each frame individually, do not achieve satisfactory results. The temporal video restoration approaches reduce the noise between frames using the motion compensation [109], but can generate blurring and dragging effects on the moving regions. The spatio-temporal denoising, or 3D denoising, techniques combine the spatial and temporal video restoration [110].

The AWGN video filtering techniques represent some extensions of the 2D image filters for Gaussian noise. They include 3D filters like the 3D Gaussian filter [111] and 3D moving average filter, the 3D patch-based filters, such as the 3D Non-Local Means (NLM) filter [112] and the spatio-temporal VBM3D (Video Block-Matching and 3D filtering) [113], tensor-based denoising schemes [114], deep learning-based filters [115] and 3D anisotropic diffusion-based models representing volumetric extensions of the Perona–Malik scheme [116]. The filtering techniques that remove the often encountered mixed Poisson–Gaussian noise from video sequences include adapted 3D NLM filters [117], improved pixel similarity weighted frame averaging (PSWFA)-based models [118], tensor-based methods [97], low rank and sparse minimization-based methods [119] and deep neural networks [120].

We have also developed spatio-temporal AWGN and mixed noise reduction techniques for video sequences using nonlinear PDE-based models. An anisotropic diffusion-based mixed noise removal approach is introduced in [121]. It is based on the following nonlinear second-order reaction–diffusion filtering model:

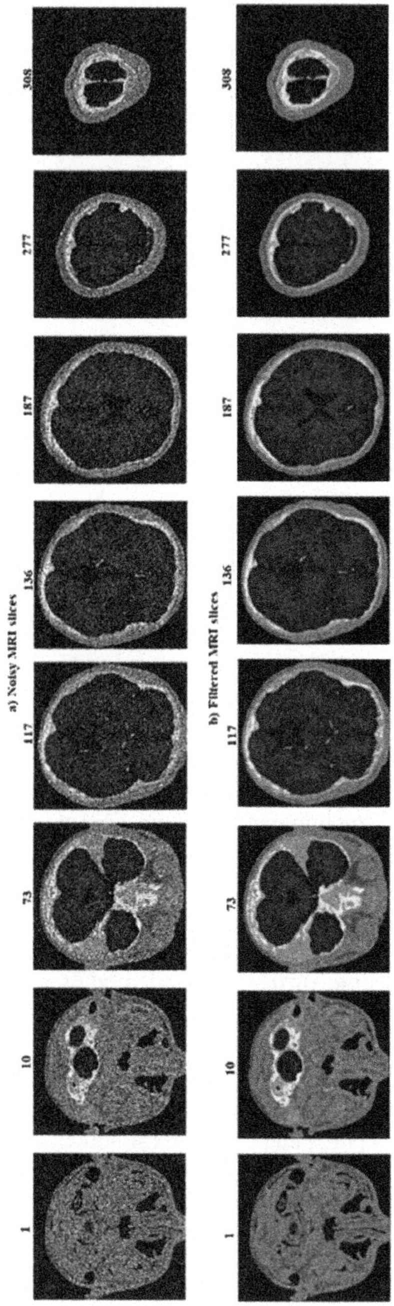

Fig. 2.21 MRI data restoration example

Table 2.12 MRI filtering method comparison

Mixed noise removal technique	PSNR score (dB)
Vector-valued diffusion-based filter	27.5342
3D median filter	24.9663
3D Gaussian filter + VST	25.4528
3D non-local mean (NLM) filter	27.2152
TV-based Poisson–Gaussian noise filter	26.5337
TV-ROF denoising + VST	25.4713

$$\begin{cases} \frac{\partial v}{\partial t} - \alpha \nabla \cdot (\eta(\|\nabla v\|)\nabla v_\sigma)\psi\left(\frac{|\nabla v|}{|\nabla^2 v|}\right) + (v - v_0)\left(\frac{\varphi}{\|v\| + \beta} + \xi\right) = 0 \\ v(x, y, z, 0) = v_0(x, y, z), \quad \forall (x,y, z) \in \Omega \\ \frac{\partial v}{\partial \vec{n}}(x, y, z, t) = 0, \quad \forall (x,y, z) \in \partial\Omega, \quad t \in (0,T] \end{cases} \tag{2.90}$$

where $v : \Omega \times (0, T] \rightarrow R$ is the analyzed video, $\lambda \in [1, 2), \varphi, \xi \in (0, 1], \beta \in (0, 0.6)$, G_σ is the 3D Gaussian filter kernel, $v_\sigma = v * G_\sigma$, the diffusivity conductance function $\eta(s) = \gamma \sqrt[r+1]{\frac{\lambda}{\log_{10}|\varphi s^r + \zeta|}}$, the function $\psi(s) = \left(\frac{\varepsilon s^r + \varsigma}{\beta}\right)^{\frac{1}{r+1}}$ and the parameters $\gamma \geq 1, \varphi \in (0, 1], \lambda, \zeta \geq 4, r \geq 2, \varepsilon, \varsigma \in [1, 5)$ [121].

The nonlinear parabolic PDE model (2.90) is non-variational, since it cannot be achieved from any energy functional minimization. It reduces successfully both signal-independent and signal-dependent components of the noise, given the term $(v - v_0)\left(\frac{\varphi}{\|v\|+\beta} + \xi\right)$, which is related to AWGN and quantum noise distributions. This nonlinear 3D reaction–diffusion process filters successfully both the spatial and temporal movie domains. The temporal frame correlation is also taken into account. The considered diffusion includes the pixels having the same location in each frame. If there is motion between consecutive frames, and thus the corresponding pixels are displaced, then the 3D gradient's magnitude, $\|\nabla v\| = \sqrt{\left(\frac{\partial v}{\partial x}\right)^2 + \left(\frac{\partial v}{\partial y}\right)^2 + \left(\frac{\partial v}{\partial z}\right)^2}$, increases and diffusivity conductance $\eta(\|\nabla v\|)$ decreases until there is no diffusion. So, the video denoising process stops whenever it reaches the edges of a moving object.

This 3D anisotropic diffusion model is also well-posed, since it admits a unique variational solution that is determined numerically using finite difference method. The PDE can be written as $\frac{\partial v}{\partial t} + (v - v_0)\left(\frac{\varphi}{\|v\|+\beta} + \xi\right) = \alpha\, div(\eta(\|\nabla v\|)\nabla v_\sigma)\psi\left(\frac{|\nabla v|}{|\nabla^2 v|}\right)$, so its left term of is approximated using the central differences [121] as:

$$\frac{v_{i,j,k}^{n+\Delta t}-v_{i,j,k}^{n}}{\Delta t}+\left(v_{i,j,k}^{n}-v_{i,j,k}^{0}\right)\left(\frac{\varphi}{\left|v_{i,j,k}^{n}\right|+\beta}+\xi\right)=v_{i,j,k}^{n+\Delta t}\frac{1}{\Delta t}$$
$$-v_{i,j,k}^{n}\left(\frac{1}{\Delta t}-\frac{\varphi}{\left|v_{i,j,k}^{n}\right|+\beta}-\xi\right)$$
$$-v_{i,j,k}^{0}\left(\frac{\varphi}{\left|v_{i,j,k}^{n}\right|+\beta}+\xi\right) \tag{2.91}$$

Then $\psi\left(\frac{|\nabla v|}{|\nabla^2 v|}\right)$ is discretized as $\psi\left(\frac{\left\|\nabla v_{i,j,k}^{n}\right\|}{\left\|\Delta v_{i,j,k}^{n}\right\|}\right)=\psi\left(\frac{\sqrt{\left(\frac{v_{i+h,j,k}^{n}-v_{i-h,j,k}^{n}}{2h}\right)^2+\left(\frac{v_{i,j+h,k}^{n}-v_{i,j-h,k}^{n}}{2h}\right)^2+\left(\frac{v_{i,j,k+h}^{n}-v_{i,j,k-h}^{n}}{2h}\right)^2}}{\frac{v_{i+h,j,k}^{n}+v_{i-h,j,k}^{n}+v_{i,j+h}^{n}+v_{i,j-h}^{n}+v_{i,j,k+h}^{n}+v_{i,j,k-h}^{n}-6v_{i,j,k}}{h^2}}\right)$ and the divergence-based term is approximated as $\tau\sum_{q\in N(\rho)}\eta\left(\left\|\nabla v_{\rho,q}^{n}\right\|\right)\nabla(v*G_\sigma)_{\rho,q}^{n}$, where $\tau\in(0,1]$, $N(\rho)$ is the 6-neighborhood of the pixel with the coordinates $\rho=\left[i,j,k\right]$ and the gradient magnitude in a particular direction is determined as $\nabla v_{\rho,q}^{n}=v_q^n-v_\rho^n$.

One sets $h=1$ and $\Delta t=1$, and by combining all these discretization results, obtains the next explicit iterative numerical approximation algorithm:

$$v_{i,j,k}^{n+1}=v_{i,j,k}^{n}\left(1-\frac{\varphi}{\left|v_{i,j,k}^{n}\right|+\beta}-\xi\right)+v_{i,j,k}^{0}\left(\frac{\varphi}{\left|v_{i,j,k}^{n}\right|+\beta}+\xi\right)$$
$$+\alpha\tau\psi\left(\frac{\sqrt{\left(v_{i+1,j,k}^{n}-v_{i-1,j,k}^{n}\right)^2+\left(v_{i,j+1,k}^{n}-v_{i,j-1,k}^{n}\right)^2+\left(v_{i,j,k+1}^{n}-v_{i,j,k-1}^{n}\right)^2}}{2\left(v_{i+h,j,k}^{n}+v_{i-h,j,k}^{n}+v_{i,j+h}^{n}+v_{i,j-h}^{n}+v_{i,j,k+h}^{n}+v_{i,j,k-h}^{n}-6v_{i,j,k}\right)}\right)$$
$$\sum_{q\in N(\rho)}\eta\left(\left\|\nabla v_{\rho,q}^{n}\right\|\right)\nabla(v*G_\sigma)_{\rho,q}^{n} \tag{2.92}$$

Table 2.13 Average PSNR scores achieved by the video filters

Video restoration technique	Average PSNR (dB)
Nonlinear reaction diffusion-based video filter	29.4168
VBM3D filter + VST	29.2307
3D non-local mean (NLM) filter	28.5573
TV-based mixed noise removal model	27.1628
3D Gaussian filter + VST	25.7604

The numerical solving scheme (2.92) is stable and consistent to the second-order reaction–diffusion model (2.88) and converges fast to its numerical solution. It was applied successfully in our movie restoration simulations that were performed on voluminous video datasets [121]. This numerical algorithm removes both signal-independent and signal-dependent components of the noise mixtures from contaminated videos, overcoming unintended effects like blurring, staircasing and flickering. Since it deals successfully with the inter-frame correlation issue, it preserves properly the essential video details both along and across the frames. This spatio-temporal video denoising method outperforms many other mixed noise removal approaches. Method comparison results are provided in Table 2.13 and a denoising example involving several frames of a traffic video is described in Fig. 2.22 [121].

We also introduced a video filtering approach for additive Gaussian noise in [122]. It is based on a fourth-order nonlinear PDE-based model having the form:

$$\begin{cases} \frac{\partial v}{\partial t} + \lambda\delta\left(\frac{\|\nabla^2 v\| + \|\nabla v\|}{2}\right)\nabla^2\left(\psi\left(\left\|\nabla^2 v\right\|\right)\Delta v_\sigma\right) - \alpha(v - v_0) = 0, \\ v(x, y, z, 0) = v_0(x, y, z), \quad \forall(x,y,z) \in \Omega \\ v(x, y, z, t) = 0, \quad \forall(x,y,z) \in \partial\Omega, \quad t \in (0,T) \end{cases} \tag{2.93}$$

The discretization and mathematical treatment of this well-posed 3D anisotropic diffusion-based filtering model are described in [122]. The simulation and method comparison results that are also presented there illustrate its effectiveness [122].

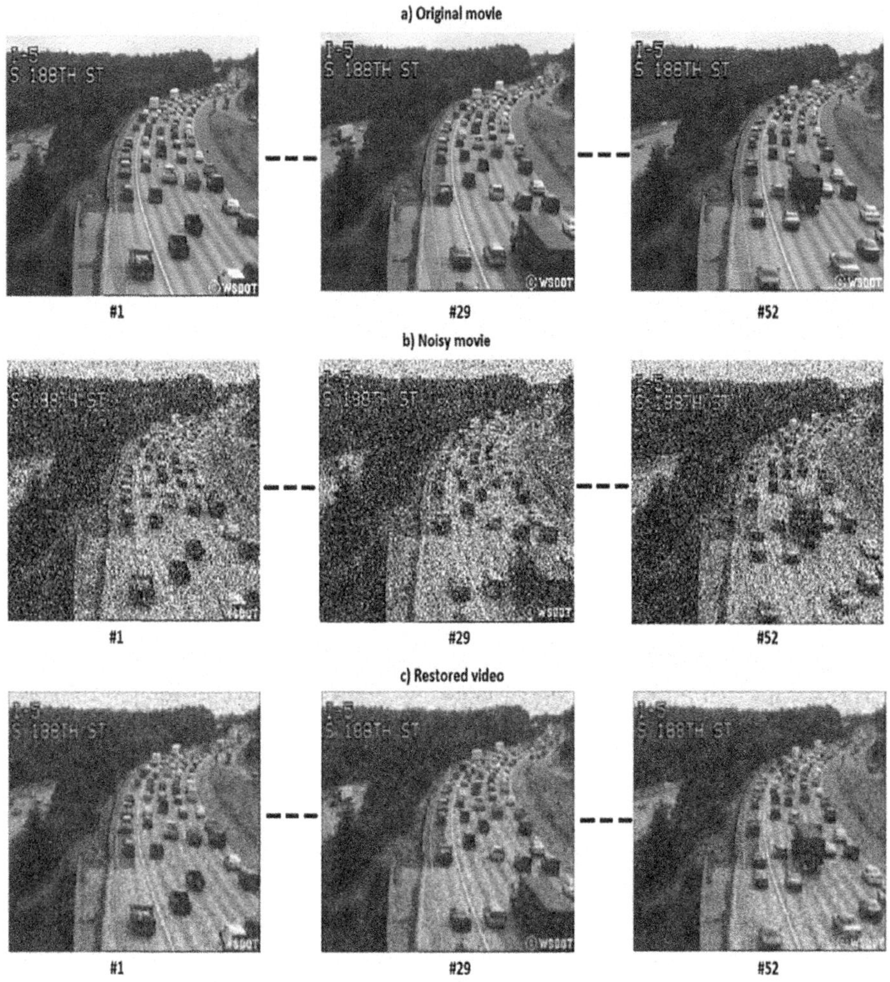

Fig. 2.22 Mixed noise removal from a traffic video

References

1. L. G. Shapiro, G. C. Stockman, *Computer Vision*, Prentice Hall, 2001.
2. R. Gonzalez, R. Woods, *Digital Image Processing*. Prentice Hall, 2nd edition, 2001.
3. H. Mehrer, *Diffusion in Solids – Fundamentals, Methods, Materials, Diffusion controlled Processes*, Springer, 2007.
4. J. Weickert, *Anisotropic Diffusion in Image Processing*, European Consortium for Mathematics in Industry. B. G. Teubner, Stuttgart, Germany, 1998.
5. M. N. Ozisik, *Heat Conduction*, 2nd edition (NY, John Wiley & Sons), 1993.
6. T. Barbu, Novel linear image denoising approach based on a modified Gaussian filter kernel, *Numerical Functional Analysis and Optimization*, Vol. 33, Issue 11, pp. 1269–1279, 2012.
7. P. Perona, J. Malik, Scale-space and edge detection using anisotropic diffusion, *Proc. of IEEE Computer Society Workshop on Computer Vision*, 16–22, nov. 1987.

8. G. Aubert, P. Kornprobst, *Mathematical problems in image processing: partial differential equations and the calculus of variations*, Vol. 147, Springer Science & Business Media, 2006.
9. L. Rudin, S. Osher, E. Fatemi, Nonlinear total variation based noise removal algorithms, *Physica D: Nonlinear Phenomena*, 60 (1), pp. 259–268, 1992.
10. P. Charbonnier, L. Blanc-Feraud, G. Aubert, M. Barlaud, Two deterministic half-quadratic regularization algorithms for computed imaging, *Proc. IEEE International Conference on Image Processing*, vol. 2, pp. 168–172, Austin, TX, IEEE Comp. Society Press, 1994.
11. J. Weickert et al., Efficient and reliable schemes for nonlinear diffusion filtering, *IEEE Transactions in Image Processing*, vol. 7, pp. 398–410, March 1998.
12. M. Black, G. Shapiro, D. Marimont, D. Heeger, Robust anisotropic diffusion, *IEEE Trans. Image Processing* 7, 3, pp. 421–432, 1998.
13. F. Voci, S. Eiho, N, Sugimoto, H. Sekiguchi, Estimating the gradient threshold in the Perona–Malik equation, *IEEE Signal Processing Magazine* 21 (3), pp. 39–46, 2004.
14. F. Catte, P. L. Lions, J. M. Morel, T. Coll, Image selective smoothing and edge detection by nonlinear diffusion, *SIAM Journal of Numerical Analysis*, Vol. 29, pp. 182–193, 1992.
15. J. Kacur, K. Mikula, Solution of nonlinear diffusion appearing in image smoothing and edge detection, *Appl. Num. Math.*, Vol. 17, 47–59, 1995.
16. F. Torkamani-Azar, K. E. Tait, Image recovery using the anisotropic diffusion equation, *IEEE Transactions on Image Processing*, Vol. 5, pp. 1573–1578, 1996.
17. H. Yu, C. S. Chua, GVF-based anisotropic diffusion models, *IEEE Transactions on Image Processing*, vol.15, no. 6, pp. 1517–1524, 2006.
18. O. Ghita, P. F. Whelan, A new GVF-based image enhancement formulation for use in the presence of mixed noise, *Pattern Recognition*, vol. 43, pp. 2646–2658, 2010.
19. Y. Wang, Y. Jia, External force for active contours: gradient vector convolution, *Pacific Rim International Conference on Artificial Intelligence*, pp. 466–472, 2008.
20. Y. Wang, W. Ren, H. Wang, Anisotropic second and fourth order diffusion models based on convolutional virtual electric field for image denoising, *Computers & Mathematics with Applications*, vol. 66, no. 10, pp. 1729–1742, 2013.
21. T. Barbu, A. Favini, Rigorous mathematical investigation of a nonlinear anisotropic diffusion-based image restoration model, *Electronic Journal of Differential Equations*, Vol. 2014, No. 129, pp. 1–9, 2014.
22. T. Barbu, *Novel Diffusion-Based Models for Image Restoration and Interpolation*, Book Series: Signals and Communication Technology, Springer International Publishing, 2019.
23. T. Barbu, A. Ciobanu, C. Niţă, Nonlinear second-order partial differential equation-based image smoothing technique, *Mem. Sci. Sect. Rom. Acad.*, pp. 7–14, Tome XXXIX, 2016.
24. T. Barbu, A Nonlinear Parabolic Partial Differential Equation Model for Image Enhancement, *International Journal of Computer and Information Engineering*, Vol. 3, No. 8, 2016.
25. L. Alvarez, P. L. Lions, J. M. Morel, Image selective smoothing and edge detection by nonlinear diffusion II, *SIAM Journal of Numerical Analysis*, 29, pp. 845–866, 1992.
26. C.A. Segall, S. T. Acton, Morphological anisotropic diffusion, *Proceedings of the 1997 IEEE International Conference on Image Processing*, Santa Barbara, CA, pp. 348–351, Oct. 26–29, 1997.
27. G. Gilboa, Y. Y. Zeevi, N. A. Sochen, Complex diffusion processes for image filtering, *Lecture Notes in Computer Science*, vol. 2106, pp. 299–307, 2001.
28. J. Weickert, Coherence-enhancing diffusion filtering, *Int. J. Comput. Vis.*, Vol. 31, No. 2/3, pp. 111–127, 1999.
29. M. Hazawinkel, *Variational calculus*, Encyclopedia of Mathematics, Springer, 2001.
30. P. Getreuer, Rudin–Osher–Fatemi Total Variation Denoising using Split Bregman, *Image Processing On Line*, 2012.
31. C. A. Micchelli, L. Shen, Y. Xu, X. Zeng, Proximity algorithms for the L1/TV image denoising model, *Advances in Computational Mathematics*, 38 (2), pp. 401–426, 2013.
32. Q. Chen, P. Montesinos, Q. Sun, P. Heng, D. Xia, Adaptive total variation denoising based on difference curvature, *Image Vision Computing*, 28 (3), pp. 298–306, 2010.

33. S. Huang, S. Wan, A total variation denoising method based on median filter and phase consistency, *Sensing and Imaging*, pp. 1–5, 2020.
34. J. Yan, W.-S. Lu, Image denoising by generalized total variation regularization and least squares fidelity, *Multidimensional Systems and Signal Processing*, vol. 26, no. 1, pp. 243–266, 2015.
35. T. Barbu, C. Morosanu, Image Restoration using a Nonlinear Second-order Parabolic PDE-based Scheme, *Analele St. ale Universitatii Ovidius Constanta*, S. Mat, Vol. XXV, 1, pp. 33–48, 2017.
36. G. Arfken, The Method of Steepest Descents, §7.4 in *Mathematical Methods for Physicists, 3rd ed.* Orlando, FL: Academic Press, pp. 428–436, 1985.
37. T. Barbu, A Novel Variational PDE Technique for Image Denoising, *Lecture Notes in Computer Science* (*Proc. of ICONIP 2013*, part III, Daegu, Korea, Nov. 3–7, 2013), Vol. 8228, pp. 501–508, Springer-Verlag Berlin Heidelberg, M. Lee et al. (Eds.), 2013.
38. T. Barbu, Nonlinear PDE Model for Image Restoration using Second-Order Hyperbolic Equations, *Numerical Functional Analysis and Optimization*, Vol. 36, Issue 11, pp. 1375–1387, Nov. 2015.
39. Y. L. You, M. Kaveh, Fourth-order partial differential equations for noise removal, *IEEE Transactions on Image Processing*, 9, pp. 1723–1730, 2000.
40. H. Wang, Y. Wang, W. Ren, Image Denoising Using Anisotropic Second and Fourth Order Diffusions Based on Gradient Vector Convolution, *Computer Science and Information Systems*, 9, pp. 1493–1512, 2012.
41. C. H. Lai, X. Y. Liu, K. A. Pericleous, A fourth-order PDE denoising model with an adaptive relaxation method, *International Journal of Computer Mathematics*, 92 (3), pp. 608–622, 2015.
42. M. Lysaker, A. Lundervold, X. C. Tai, Noise removal using fourth-order partial differential equation with applications to medical magnetic resonance images in space and time, *IEEE Transactions on Image Processing* 12, pp. 1579–1590, 2003.
43. T. F. Chan, A. Marquina, P. Mulet, High-order total variation based image restoration, *SIAM J. Sci. Comput.* 22, pp. 503–516, 2000.
44. S. Kim, H. Lim, Fourth-order partial differential equations for effective image denoising, *Electronic Journal Differ. Equ. Conf.*, 17, pp. 107–121, 2009.
45. F. Li, C. M. Shen, J. Fan, C. Shen, Image restoration combining a total variational filter and a fourth-order filter, *Journal of Visual Communication and Image Representation* 18 (4), pp. 322–330, 2007.
46. T. Liu, Z. Xiang, Image restoration combining the second order and fourth-order PDEs, *Mathematical Problems in Engineering*, vol. 2013, Article ID 743891, 7 pages, 2013.
47. O. M. Lysaker, X.-C. Tai, Iterative image restoration combining total variation minimization and a second-order functional, *International Journal of Computer Vision*, 66, pp. 5–18, 2006.
48. V. S. Prasath, P. Kalavathi, Mixed Noise Removal Using Hybrid Fourth Order Mean Curvature Motion, *Advances in Signal Processing and Intelligent Recognition Systems*, pp. 625–632, Springer International Publishing, 2016.
49. T. Barbu, I. Munteanu, A Nonlinear Fourth-order Diffusion-based Model for Image Denoising and Restoration, *Proceedings of the Romanian Academy*, Series A, Vol. 18, Issue 2, pp. 108–115, 2017.
50. T. Barbu, PDE-based Restoration Model using Nonlinear Second and Fourth Order Diffusions, *Proceedings of the Romanian Academy*, Series A, Vol. 16, No. 2, pp. 138–146, April-June 2015.
51. F. A. Haight, *Handbook of the Poisson Distribution*, NY: John Wiley & Sons. 1967.
52. A. Buades, B. Coll, J.-M. Morel, A review of image denoising algorithms, with a new one, *Multiscale Model. Simul.* 4(2), pp. 490–530. 2005.
53. K. V. Thakur, O. H. Damodare, A. M. Sapkal, Poisson Noise Reducing Bilateral Filter. *Procedia Computer Science*, 79, pp. 861–865, 2016.
54. B. Zhang, J. M. Fadili, J. Starck, Multi-scale Poisson count image denoising, *Proceedings of IEEE ICASSP'2006*, Toulouse, France, 2006.

55. R. Nowak, M. Figueiredo, *Unsupervised progressive parsing of poisson fields using minimum description length criteria*, Proc. of Intl. Conf. on Image Proc., ICIP 99, vol. 2, pp. 26–30, 1999.
56. B. Zhang, J. M. Fadili, J. Starck, Wavelets, ridgelets, and curvelets for Poisson noise removal. *IEEE Trans. Image Processing*, vol. 17, pp. 1093–1108, 2008.
57. T. Le, R. Chartrand, T. J. Asaki, A variational approach to reconstructing images corrupted by Poisson noise, *J. Math. Imaging Vis.*, 27 (3), pp. 257–263, 2007.
58. W. Zhou, Q. Li, Poisson noise removal scheme based on fourth-order PDE by alternating minimization algorithm, *Abstr. Appl. Anal.*, vol. 2012, Nov. 2012.
59. A. Sawatzky, C. Brune, J. Muller, M. Burger, Total variation processing of images with poisson statistics, *Computer Analysis of Images and Patterns*, Springer, pp. 533–540, 2009.
60. M. R. Chowdhury, J. Zhang, J. Qin, Y. Lou, Poisson image denoising based on fractional-order total variation, *Inverse Problems & Imaging*, 14 (1), pp. 77–96, 2020.
61. X. Liu, L. Huang, Total bounded variation-based Poissonian images recovery by split Bregman iteration, *Mathematical Methods in the Applied Sciences*, 35(5), pp. 520–529, 2012.
62. W. Zhou, Q. Li, Poisson noise removal scheme based on fourth-order PDE by alternating minimization algorithm, *Abstract and Applied Analysis*, Vol. 2012, Hindawi, 2012.
63. M. Ma, J. Yang, Poisson image restoration via an adaptive Euler's elastica regularization, *Journal of Intelligent & Fuzzy Systems*, pp. 1–16, 2023.
64. J. Darbon, T. Meng, E. Resmerita, On Hamilton–Jacobi PDEs and image denoising models with certain nonadditive noise, *Journal of Mathematical Imaging and Vision*, 64 (4), pp. 408–441, 2022.
65. T. Barbu, Second-Order Anisotropic Diffusion-Based Technique for Poisson Noise Removal, *IFAC PapersOnLine*, Elsevier Ltd., Volume 52, Issue 18, pp. 174–178, 2019.
66. T. Barbu, A Nonlinear Second-order Hyperbolic PDE-based Photon-limited Medical Microscopy Image Restoration Technique, *Romanian Journal of Information Science and Technology* (*ROMJIST*), Volume 23, Number S, pp. S67–S76, 2020.
67. T. Barbu, Variational Quantum Denoising Technique for Medical Images, *8th International Conference on e-Health and Bioengineering, EHB 2020*, Iaşi, Romania, pp. 1–4, 29–30 Oct. 2020, IEEE.
68. T. Barbu, Photon-limited Image Restoration Using a Well-posed Nonlinear Fourth-order Hyperbolic PDE-based Model, *Proceedings of the Romanian Academy Series A*, 21(3), pp. 205–212, 2020.
69. J. - L. Lions, W. Strauss, Some non-linear evolution equations, *Bulletin de la Société Mathématique de France*, 93, pp. 43–96, 1965.
70. F. Luisier, T. Blu, M. Unser, Image denoising in mixed Poisson-Gaussian noise, *IEEE Trans. Image Process.* 20, pp. 696–708, 2010.
71. F. Murtagh, J. Starck, A. Bijaoui, Image restoration with noise suppression using a multi resolution support, *Astron. Astrophys. Supplement Series*, vol. 112, pp. 179–189, 1995.
72. S. Yang, B. U. Lee, Poisson-Gaussian noise reduction using the hidden Markov model in contourlet domain for fluorescence microscopy images, *PLOS ONE*, vol. 10, no. 9, pp. 1–19, 2015.
73. C. T. Pham, T. Tran, G. Gamard, An Efficient Total Variation Minimization Method for Image Restoration, *INFORMATICA*, Vol. 31, No. 3, pp. 539–560, 2020.
74. D. N. H. Thanh, S. D. Dvoenko, A method of total variation to remove the mixed Poisson-Gaussian noise, *Pattern Recognition and Image Analysis*, vol. 26, no. 2, pp. 285–293, 2016.
75. C. T. Pham, T. Thao, P. Khoa, D. Sang, P. Tuan, N. Hieu, An adaptive algorithm for restoring image corrupted by mixed noise, *Cybernetics and Physics*, 8 (2), pp. 73–82, 2019.
76. A. Chambolle, An algorithm for total variation minimization and applications, *Journal of Mathematical Imaging and Vision*, 20, pp. 89–97, 2004.
77. A. Lanza, S. Morigi, F. Sgallari, Y. Wen, Image restoration with Poisson-Gaussian mixed noise, *Computer Methods in Biomechanics and Biomedical Eng.: Imaging & Vis.*, 2 (1), pp. 12–24, 2014.

78. C. T. Pham, T. Thao, T. Cong, V. Hoang, Second-order total generalized variation based model for restoring images with mixed Poisson-Gaussian noise, *Information & Control Systems*, 111 (2), 2021.
79. T. Barbu, Novel variational technique for mixed Poisson-Gaussian noise filtering, *15th IEEE International Symposium on Signals, Circuits and Systems, ISSCS 2021*, Iasi, Romania, July 2021.
80. T. Barbu, Mixed Noise Removal Framework using a Nonlinear Fourth-order PDE-based Model, *Applied Mathematics & Optimization*, 84, August 2021, pp. 1865–1876, Springer US.
81. N. Hagen, M. W. Kudenov, Review of snapshot spectral imaging technologies, *Optical Engineering*, 52 (9), 2013.
82. P. Reddy, S. Pawar, Multispectral image denoising methods: A literature review, *Materials Today: Proceedings*, 33, pp. 4666–4670, 2020.
83. C. I. Chang, *Hyperspectral Imaging: Techniques for Spectral Detection and Classification*, Springer Science & Business Media, 31 July 2003.
84. C. Brown, C. Harder, *The ArcGIS Imagery Book: New View. New Vision*, Esri Press., 2016.
85. J. Mairal, M. Elad, G. Sapiro, Sparse representation for colour image restoration, *IEEE Trans. Image Process.* 17 (1), pp. 53–69, Dec. 2007.
86. Y. Qian, Y. Shen, M. Ye, Q. Wang, 3-D nonlocal means filter with noise estimation for hyperspectral imagery denoising, *IEEE Intl. Geoscience and Rem. Sensing Symp.*, pp. 1345–1348, 2012.
87. G. Papari, N. Idowu, T. Varslot, Fast Bilateral Filtering for Denoising Large 3D Images, *IEEE Transactions on Image Processing*, vol. 26, no. 1, pp. 251–261, Jan. 2017.
88. Y. Peng et al, Decomposable nonlocal tensor dictionary learning for multispectral image denoising, *Proc. of IEEE Conf. on Comp. Vision and Pattern Recognition*, pp. 2949–2956, 2014.
89. S. Suresh, S. Lal, C. Chen, T. Celik, Multispectral satellite image denoising via adaptive cuckoo search-based wiener filter, *IEEE Trans. Geosci. Remote Sens.* 56 (8), pp. 4334–4345, 2018.
90. K. Dabov, A. Foi, V. Katkovnik, K. Egiazarian, Image denoising by sparse 3d transform-domain collaborative filtering, *IEEE Trans. Image Processing*, 16 (8), pp. 2080–2094, 2007.
91. U. Ojha, A. Garg, Denoising High Resolution Multispectral Images Using Deep Learning Approach, *15th IEEE Intl. Conf. on Mach. Learning and App. (ICMLA)*, Anaheim, US, pp. 871–875, 2016.
92. M. Lee, S. Park, S. Lee, M. Kang, Motion-adaptive 3D nonlocal means filter based on stochastic distance for low-dose X-ray fluoroscopy, *Biomedical Signal Proc. and Control*, 38, pp. 74–85, 2017.
93. M. Mäkitalo, A.Foi, Poisson-Gaussian denoising using the exact unbiased inverse of the generalized Anscombe transformation, *IEEE International Conference on Acoustics, Speech and Signal Processing* (ICASSP), Kyoto, Japan, pp. 1081–1084, 2012.
94. C. Zou, Y. Xia, Restoration of hyperspectral image contaminated by poisson noise using spectral unmixing, *Neurocomputing*, 275, pp. 430–437, 2018.
95. A. Mansouri, F. Deger, M. Pedersen, J. Hardeberg, Y. Voisin, An adaptive spatial–spectral total variation approach for Poisson noise removal in hyperspectral images, *Signal, Image and Video Processing*, 10, pp. 447–454, 2016.
96. Y. Zhao, J. Yang, J. Chan, Hyperspectral imagery super-resolution by spatial–spectral joint nonlocal similarity, *IEEE J. of Sel. Topics in Appl. Earth Obs.and Remote Sens*, 7(6), pp. 2671–2679, 2013.
97. W. Dong, T. Huang, G. Shi, Y. Ma, X. Li, Robust tensor approximation with Laplacian scale mixture modeling for multiframe image and video denoising, *IEEE Journal of Selected Topics in Signal Processing*, 12 (6), pp. 1435–1448, 2018.
98. Y. C. Heo, K. Kim, Y. Lee, Image denoising using non-local means (NLM) approach in magnetic resonance (MR) imaging: a systematic review, *Applied Sciences*, 10 (20), 7028, 2020.

99. M. Lennon, G. Mercier, L. H.-Moy, Nonlinear filtering of hyperspectral images with anisotropic diffusion, *IEEE Intl. Geoscience and Remote Sensing Symposium*, Vol. 4, pp. 2477–2479, 2002.
100. R. Whitaker, G. Gerig, Vector-valued diffusion, *Geometry-driven diffusion in computer vision*, pp. 93–134, Dordrecht: Springer Netherlands, 1994.
101. D. Tschumperle, R. Deriche, Vector-valued image regularization with PDEs: A common frame work for different applications, *IEEE trans. on pat. anal. and mach. intell.*, 27(4), pp. 506–517, 2005.
102. D. Tschumperle, R. Deriche, Diffusion PDEs on vector-valued images, *IEEE Signal Processing Magazine*, 19 (5), pp. 16–25, 2002.
103. T. Barbu, Spectral Vector-valued Image Restoration using a Hyperbolic Partial Differential Equation-based Filter, *12th Mediterranean Conference on Embedded Computing, MECO 2023*, Budva, Montenegro, June 6–10, 2023. IEEE.
104. T. Barbu, Vector-valued Fourth-order Reaction-diffusion based Photon-limited Multi-channel Image Filtering, *16th Intl. Symp. on Signals, Circ and Sys., ISSCS'23*, Iasi, Romania, July 13–14, 2023.
105. T. Barbu, Multispectral Image Restoration using a Vector-valued Reaction-diffusion based Mixed Noise Removal Technique, *ISPRS Annals of the Photogrammetry, Remote Sensing and Spatial Information Sciences* X - 1 / W1 - 2023, pp. 351–356, 2023.
106. T. Barbu, Hyperspectral Data Volume Restoration using Nonlinear Vector-valued Reaction-diffusion based Filtering Scheme, *CoDIT 24–10th Intl. Conf. on Ctrl, Decision and Inf. Tech.*, 2024.
107. T. Barbu, Multimodal Magnetic Resonance Image Restoration using a Vector-valued Anisotropic Diffusion-based Mixed Noise Filtering Technique, *11th International Conference on e-Health and Bioengineering, EHB 2023*, Bucureşti, Romania, pp. 1–4, 9–10 Nov. 2023.
108. Z. Bi, *Comprehensive Survey on Video Denoise Methods*, 2022.
109. C. Cheng, C. Chen, Y. Chen, L. Chen, Analysis and VLSI architecture of update step in motion-compensated temporal filtering, *IEEE International Symp. on Circuits and Systems (ISCAS)*, 2006.
110. M. Szczepanski, Spatio-Temporal Filters in Video Stream Processing, *Computer Recognition Systems 4. Advances in Intelligent and Soft Computing*, vol. 95, Springer, Berlin, Heidelberg, 2011.
111. O. Wirjadi, T. Breuel, Approximate separable 3D anisotropic Gauss filter, *IEEE International Conference on Image Processing 2005*, Genova, Italy, 2005, pp. II–149.
112. R. A. Ali, R. C. Hardie, Recursive non-local means filter for video denoising, *EURASIP Journal on Image and Video Processing*, pp. 1–11, 2017.
113. T. Ehret, P. Arias, Implementation of VBM3D and Some Variants, *Image Processing On Line*, 11, pp. 374–395, 2021.
114. L. Gui, G. Cui, Q. Zhao, D. Wang, A. Cichocki, J. Cao, Video denoising using low rank tensor decomposition, *9th International Conf. on Machine Vision-ICMV*, Vol. 10341, pp. 162–166, 2016.
115. M. Tassano, J. Delon, T. Veit, Fastdvdnet: Towards real-time deep video denoising without flow estimation, *Proc.of IEEE/CVF conf. on computer vision and pattern rec.*, pp. 1354–1363, 2020.
116. S. Lee, M. Kang, Spatio-temporal video filtering algorithm based on 3-D anisotropic diffusion equation, *International Conference on Image Processing. ICIP98*, Vol. 2, pp. 447–450, 1998. IEEE.
117. R. Almahdi, R.Hardie, Recursive non-local means filter for video denoising with Poisson-Gaussian noise, *IEEE Nat. Aerospace and Electronics Conf. and Ohio Innovation Summit*, pp. 318–322, 2016.
118. X. Tan, Y. Liu, C. Zuo, M. Zhang, A real-time video denoising algorithm with FPGA implementation for Poisson–Gaussian noise, *Journal of Real-Time Image Processing*, 13, pp. 327–343, 2017.

119. W. Li, X. Yin, Y. Liu, M. Zhang, Robust video denoising for mixed Poisson, Gaussian and impulse noise, *2nd Intl. Conference on Image, Vision and Computing (ICIVC)*, Chengdu, pp. 459–462, 2017.
120. H. Zheng, T. Pang, H. Ji, Unsupervised deep video denoising with untrained network, *Proceedings of the AAAI Conference on Artificial Intelligence*, Vol. 37, No. 3, pp. 3651–3659, 2023.
121. T. Barbu, C. Moroşanu, Nonlinear Reaction-diffusion based Video Restoration Technique for Noise Mixtures, *19th IEEE Conference on Industrial Electronics and Applications, ICIEA 2024*, pp. 1–4, Kristiansand, Norway, August 5–8, 2024.
122. T. Barbu, Edge Detection Techniques using Nonlinear Diffusion – based Models, *Bulletin of the Transilvania University of Brasov. Series III*, Vol. 6 (68), No. 2, 2025.

Chapter 3
Nonlinear PDE-Based Models for Inpainting and Compression

Two closely related image processing domains based on nonlinear partial differential equations, namely image inpainting and compression are described in this chapter. The nonlinear PDE-based image reconstruction, or inpainting, field is approached in the first section. Variational and non-variational structural inpainting techniques based on nonlinear PDE models of various orders are described here. Our main contributions in this domain are also surveyed in this section.

The PDE-based image compression area is then described in the second section. It can be regarded as an application field of the image interpolation, since the PDE-based inpainting models are applied successfully to the decompression task. The most important nonlinear diffusion-based decompression schemes are surveyed first. Next, our major contributions in the nonlinear PDE-based compression domain are discussed here.

3.1 Variational and Anisotropic Diffusion Models for Structural Inpainting

The image *reconstruction*, which is also called image *interpolation*, *inpainting* or *completion*, constitutes the process of recovering the missing or highly deteriorated image areas as plausibly as possible, by using the information achieved from the surrounding regions [1]. There are three categories of reconstruction techniques: textural inpainting, structural inpainting and combined approaches. The texture-based methods can be related to texture synthesis [2] or may represent exemplar-based approaches [3]. Structure-based inpainting employs information around the damaged region to estimate isophote from coarse to fine, and diffuses it by a diffusion mechanism that uses variational and PDE-based models [1, 4].

T. Barbu, *Digital Image Processing, Analysis and Computer Vision Using Nonlinear Partial Differential Equations*, Studies in Computational Intelligence 1211,
https://doi.org/10.1007/978-3-031-89576-0_3

3.1.1 Variational Structure-Based Image Interpolation Algorithms

A variational inpainting model is based on a energy minimization having the form:

$$\min_{u\in U}\left(J(u)=R(u)+\frac{1}{2}\int_{\Omega}\lambda_D(u-u_0)^2 d\Omega\right) \tag{3.1}$$

where D is the inpainting domain whose *mask* is given by $\lambda_D=\lambda\cdot 1_{\Omega\backslash D}, \lambda>0$, $R(u)$ represents the *regularizing term* responsible for the filling process and contains certain a-priori information of image u and the fidelity term $\frac{1}{2}\int_{\Omega}\lambda_D(u-u_0)^2 d\Omega$ forces u to remain close enough to the observation u_0 outside of D [4]. Various inpainting solutions can be provided by various forms of the regularizer term. For example, Harmonic Inpainting model uses $R(u)=\int_{\Omega}\|\nabla u\|^2 dxdy$ [4].

The early energy-based inpainting schemes were based on the influential Mumford–Shah model. One of them uses the Γ-convergence approximation of the Mumford–Shah functional [5] and consists of the next minimization:

$$\min_{u\in U}\left(\frac{1}{2}\int_{\Omega}\lambda_D(u-u_0)^2 d\Omega+\frac{\gamma}{2}\int_{\Omega\backslash\Gamma}z_{\in}^2|\nabla u|^2 d\Omega\right.$$
$$\left.+\alpha\int_{\Omega\backslash\Gamma}\left(\in|\nabla z_{\in}|^2+\frac{(1-z_{\in})^2}{4\in}\right)d\Omega\right) \tag{3.2}$$

where $z_{\in}$ stands for the signature function of the edge set. This inpainting approach has a low order of complexity in terms of approximation and computation. Also, it has a fast numerical convergence and preserves the boundary sharpness. Another inpainting method, based on the Mumford–Shah–Euler model [6], has the form:

$$\min_{u}\left(\frac{1}{2}\int_{\Omega}\lambda_D(u-u_0)^2 dxdy+\frac{\gamma}{2}\int_{\Omega\backslash\Gamma}|\nabla u|^2 dxdy+\alpha\int_{\Gamma}\left(\alpha+\beta\kappa^2\right)ds\right) \tag{3.3}$$

Total Variation (TV) Inpainting, an influential interpolation model introduced by Chan and Shen [7], is characterized by the energy functional minimization:

$$\min_{u\in L^2(\Omega)}\left(J[u]=\int_{\Omega}\alpha\|\nabla u\|dxdy+\frac{\lambda_D}{2}\int_{\Omega}(u-u_0)^2 dxdy\right) \tag{3.4}$$

where $\alpha > 0$ and $\lambda_D = \lambda \cdot (1 - 1_D)$. It fills in the missing zones by minimizing the first-order total variation while keeping close to the observed image in the known regions. The coefficient α controls the denoising process: a low value for directs the filtering mostly to inpainting zone and is minimal outside of D. The next second-order nonlinear anisotropic diffusion-based model is then derived from (3.5), by applying its Euler–Lagrange equation and the steepest descent method:

$$\begin{cases} \frac{\partial u}{\partial t} = \alpha\, div\left(\frac{\nabla u}{|\nabla u|}\right) - \lambda_D(u - u_0) \\ u(0, x, y) = u_0(x, y) \end{cases}, \tag{3.5}$$

TV Inpainting works successfully for non-textured images corrupted by missing or highly damaged regions. It achieves connectivity, but not for large gaps, and may produce staircasing, since TV minimization generates piecewise constant structures. Since inpainting represents filtering with spatially-varying regularization strength λ_D, the TV regularization-based inpainting is strongly related to TV-ROF Denoising provided in (2.14)–(2.15). Given their similarity, TV Denoising with Split Bregman algorithm was adapted easily to the interpolation field [8].

Numerous variational completion techniques that improve TV Inpainting have been developed in the last decades. One of them is TV^2 Inpainting model, obtained by using the second-order total variation as regularizer term: $R(u) = \int_\Omega |\nabla^2 u| dxdy$ [9].

It outperforms TV Inpainting, achieving better and more natural reconstructions, and overcomes the staircase effect. Also, TV^2 Inpainting represents a better connectivity solution, providing an effective interpolation along large gaps [1, 9].

Some variational models combining the first- and second-order TV regularizations achieve improved interpolation results. The TV + TV^2 Inpainting technique described in [10] obtains the inpainted image u^* as following:

$$u^* = \arg\min_u \left(\frac{\lambda_D}{2} \int\limits_{\Omega \backslash D} (u - u_0)^2 dxdy + \int\limits_\Omega \alpha(x)|\nabla u| dxdy + \int\limits_\Omega \beta(x)|\nabla^2 u| dxdy \right) \tag{3.6}$$

where $\alpha(x)$ and $\beta(x)$ are 2 properly modeled spatially varying functions. This combined method provides effective filling results, outperforming both TV and TV^2 Inpainting. It interpolates successfully large inpainting regions, does not create blocky-like structures and is numerically solved using Split Bregman algorithm [10].

The Total Generalized Variation (TGV) used for image denoising was also adapted for inpainting in [11]. Other effective total variation-based completion solutions are Blind Inpainting using l_0 and TV Regularization [12] and TV Inpainting with Primal-Dual Active Set (PDAS) [13]. An influential higher-order variational inpainting framework is Euler's Elastica Inpainting model proposed by Chan and Shen [14]. It results from (3.1) applying the following regularizer:

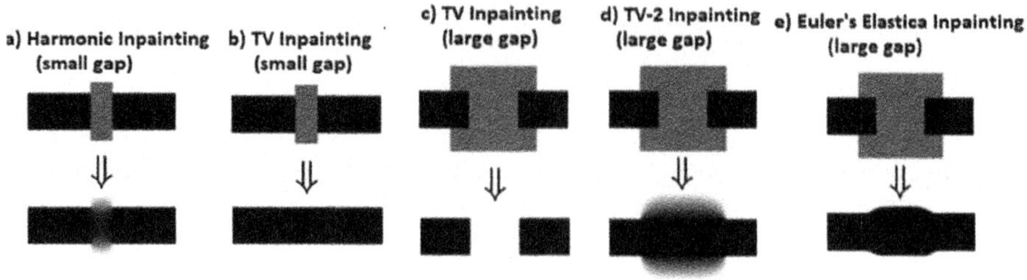

Fig. 3.1 Connectivity results of variational inpainting models

$$R(u) = \int_{\Omega} w(u)\left(\alpha + \beta\left(\nabla \cdot \frac{\nabla u}{|\nabla u|}\right)^2\right)|\nabla u| dxdy \tag{3.7}$$

where $\alpha, \beta > 0$ control its behavior and $w(u)$ is a histogram-based weighting function. Euler's Elastica Inpainting provides a proper reconstruction of the image, being able to fill-in large missing regions and working in noisy conditions too. It achieves a better connectivity than TV Inpainting, interpolating along much larger gaps [14]. The connectivities of the described inpainting models are illustrated in Fig. 3.1.

We also developed some effective variational inpainting techniques that were surveyed in [15]. Such an energy-based reconstruction approach, which is disseminated in [16], recovers the damaged image as $u_{opt} = \arg\min_u F(u)$ where

$$F(u) = \int_{\Omega} \left(\alpha\varphi_u(\|\nabla u\|) + \frac{\beta}{2}(1 - 1_{\Gamma})(u - u_0)^2\right) d\Omega \tag{3.8}$$

where $\alpha, \beta \in (0, 1)$ and its regularizer is $\varphi_u(s) = \int_0^s \tau\xi\left(\frac{\eta(u)}{\gamma \log_{10}(s+\eta(u))^2+\delta}\right)^{\frac{1}{3}} d\tau$, with $\eta(u) = \zeta \cdot median(\|\nabla u\|) + vt, \xi, \gamma, \zeta, v \in (0, 1]$ and $\delta \in [1, 5)$ [16].

By applying the Euler–Lagrange equation and gradient descent [16], one obtains:

$$\begin{cases} \frac{\partial u}{\partial t} = 2\alpha\, div(\psi_u(\|\nabla u\|)\nabla u) - \beta(1 - 1_{\Gamma}) \cdot (u - u_0) \\ u(0, x, y) = u_0 \\ u(t, x, y) = 0, \quad on\ \partial\Omega \backslash \Gamma \end{cases} \tag{3.9}$$

A rigorous mathematical treatment of the validity of the nonlinear second-order anisotropic diffusion model (3.9) was performed, its well-posedness being demonstrated [16]. This PDE-based model was then solved numerically using the finite difference method [16], the next iterative numerical approximation being obtained:

$$
\begin{aligned}
u_{i,j}^{n+1} &= u_{i,j}^{n}(1-\beta(1-1_\Gamma)) + 2\alpha\psi_u\left(\frac{\sqrt{\left(u_{i+1,j}^{n}-u_{i-1,j}^{n}\right)^2+\left(u_{i,j+1}^{n}-u_{i,j-1}^{n}\right)^2}^{2}}{2}\right)\\
&\cdot\left(u_{i+1,j}^{n}+u_{i-1,j}^{n}+u_{i,j+1}^{n}+u_{i,j-1}^{n}-4u_{i,j}^{n}\right)\\
&+2\alpha\psi_u'\left(\frac{\sqrt{\left(u_{i+1,j}^{n}-u_{i-1,j}^{n}\right)^2+\left(u_{i,j+1}^{n}-u_{i,j-1}^{n}\right)^2}^{2}}{2}\right)\\
&\cdot\frac{\left(u_{i+1,j+1}^{n}-u_{i+1,j-1}^{n}-u_{i-1,j+1}^{n}+u_{i-1,j-1}^{n}\right)\left(u_{i+1,j}^{n}-u_{i-1,j}^{n}+u_{i,j+1}^{n}-u_{i,j-1}^{n}\right)}{8}\\
&+\beta(1-1_\Gamma)u_{i,j}^{0}
\end{aligned}
\tag{3.10}
$$

where $n = 0, \ldots, N$, $I = 1, \ldots, I$ and $J = 1, \ldots, J$. The approximation algorithm is consistent to the diffusion model and converges fast to its solution [16]. This technique completes successfully the missing parts, preserving the essential details and avoiding unintended effects. See some method comparison results in Table 3.1 [16].

Another variational inpainting framework proposed by us is based on a combination of nonlinear second and fourth order diffusions [17]. It inpaints the image by minimizing the next energy cost functional based on two regularizer functions:

$$
u_{inp} = \arg\min_u \int_\Omega \left(\varphi_u^1(\|\nabla u\|) + \varphi_u^2(|\Delta u|) + \frac{\alpha(1-1_\Gamma)}{2}(u-u_0)^2\right)d\Omega \tag{3.11}
$$

with $\alpha \in (0, 1]$, $\Gamma \subset \Omega$ and the regularizers $\varphi_u^1, \varphi_u^2 : (0, \infty) \to (0, \infty)$ are

Table 3.1 Average PSNR values of several inpainting approaches

Inpainting technique	Average PSNR score (dB)
Our variational model	31.25
TV Inpainting	26.38
Harmonic Inpainting	23.54
P-M based Inpainting	26.38
Kriging Interpolation	30.38
TV with PDAS	33.96

Table 3.2 Average PSNR achieved by variational inpainting models

Inpainting method	Average PSNR (dB)
The proposed variational scheme	31.95
TV Inpainting	29.96
Harmonic Inpainting	27.32
Mumford–Shah based Inpainting	30.27
Navier–Stokes based Inpainting	31.75

$$\begin{cases} \varphi_u^1(s) = \int\limits_0^s \tau\lambda\left(\frac{\eta_u}{\beta\tau^k+\xi}\right)^{\frac{1}{3}} d\tau \\ \varphi_u^2(s) = \int\limits_0^s \tau\xi\sqrt{\frac{\eta_u}{\varepsilon\log_{10}(\eta_u+\tau)^3+\gamma}}d\tau \end{cases} \tag{3.12}$$

where $\lambda, \beta, \varepsilon \in (0, 1], \xi, \gamma \in [1, 3], k \in (0, 4), \eta_u = |\nu\mu(\|\nabla u\|) - \zeta t|$ and $\zeta, \nu \in (0, 1)$.

Next 4th-order diffusion model with boundary conditions is derived from (3.10):

$$\begin{cases} \frac{\partial u}{\partial t} = div\left(\delta_1^u(\|\nabla u\|)\nabla u\right) - \Delta\left(\delta_2^u\left(\left|\nabla^2 u\right|\right)\Delta u\right) - \alpha(1 - 1_\Gamma)(u - u_0) \\ u(t, x, y) = 0, \quad on\ \partial\Omega\backslash\Gamma \\ u(0, x, y) = u_0 \\ \frac{\partial u}{\partial \vec{n}} = 0 \end{cases} \tag{3.13}$$

where $\delta_1^u, \delta_2^u : [0, \infty) \to (0, \infty), \delta_1^u(s) = \frac{\varphi_u^{1'}(s)}{s}, \delta_2^u(s) = \frac{\varphi_u^{2'}(s)}{s}$ [17].

A finite difference-based numerical approximation of this nonlinear diffusion model is performed in [17]. The obtained fast numerical solver provides a good structural completion of the damaged images, while preserving the edges and other features and overcoming the blurring and staircasing. Well-known image collections, like the USC-SIPI database, were used in our inpainting tests. Method comparison results are provided in Table 3.2 and some inpainting examples in Fig. 3.2.

3.1.2 Non-variational PDE-Based Structural Inpainting Solutions

Numerous nonlinear PDE-based inpainting techniques introduced in the last 25 years, in both variational and non-variational form. Usually, the second-order PDE-based interpolation models have the next general form:

$$\frac{\partial u}{\partial t} + \nabla R(u) + \lambda_D(u - u_0) = 0 \text{ in } \Omega \tag{3.14}$$

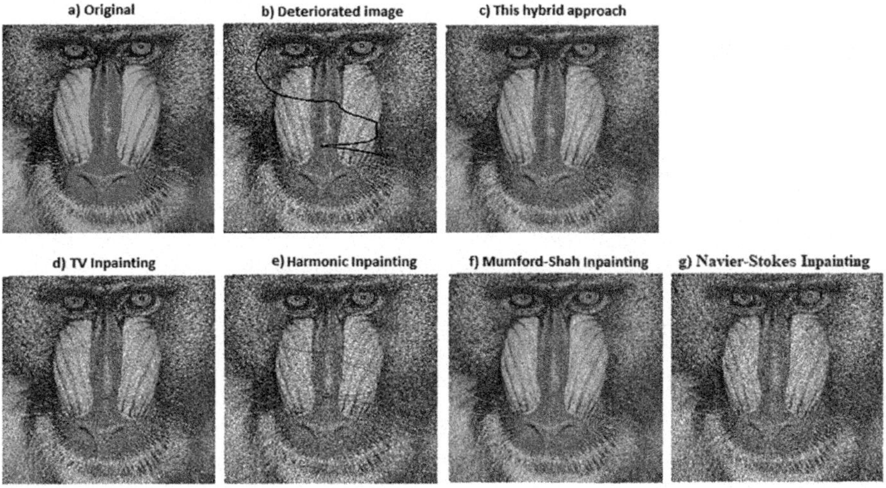

Fig. 3.2 Inpainting results of several variational approaches

where $\nabla R(u)$ is the Frechet derivative of the regularizing function. Most of them follow variational principles, since can be derived from energy-based models, but some diffusion-based inpainting schemes are non-variational. We introduced such non-variational PDE-based inpainting models in our past works.

A nonlinear parabolic PDE-based inpainting technique is disseminated in [18]. It uses the following nonlinear anisotropic diffusion equation:

$$\begin{cases} \frac{\partial u}{\partial t} - \alpha\, div(\psi_u(\|\nabla u\|)\nabla u) + \gamma(1 - 1_\Gamma)(u - u_0) + F \cdot \nabla u \\ u(0, x, y) = u_0 \\ u(t, x, y) = 0, \quad on\ \partial\Omega \backslash \Gamma \end{cases} \tag{3.15}$$

where $\alpha \in (0, 6], \gamma \in (0, 1]$ and the drift term $F(x, y) = \left(e^{-\lambda_1(x^k+y^k)}, e^{-\lambda_2(x^{k+1}+y^{k+1})}\right)$, with $k, \lambda_1, \lambda_1 > 0$. The next conductance function is proposed for this model:

$$\begin{aligned} &\psi_u : [0, \infty) \to [0, \infty) \\ &\psi_u(s) = \frac{\lambda}{\eta(u)\left|\ln\left(\frac{s}{\eta(u)}\right)\right|^k + \delta\left(\frac{s}{\eta(u)}\right)^2} \end{aligned} \tag{3.16}$$

where $\alpha, \delta \in (1, 3]$ and $\eta(u) = \left|\frac{\xi\mu(\|\nabla u\|)+\beta \cdot median(\|\nabla u\|)}{\nu}\right|$, with $\xi, \beta, \nu \in (0, 3]$. An explicit numerical approximation scheme is built for it using finite differences [18].

The proposed anisotropic diffusion models provides good structural interpolation results, outperforming some existing PDE-based schemes. See some method comparison results based on the average Norm of the Error (NE) value in Table 3.3.

Table 3.3 NE values achieved by several PDE-based methods

Inpainting approach	Norm of the error
Our PDE-based inpainting model	5.3×10^3
TV Inpainting	5.9×10^3
Harmonic Inpainting	6.4×10^3

Another structure-based inpainting technique is based on the next nonlinear second-order hyperbolic PDE-based model with boundary conditions [19]:

$$\begin{cases} \gamma \frac{\partial^2 u}{\partial t^2} + \lambda^2 \frac{\partial u}{\partial t} - div(\xi_u(|\nabla u|) \cdot \nabla u) + \beta(1 - 1_M)(u - u_0) = 0 \\ u(0, x, y) = u_0(x, y) \\ \dfrac{\partial u}{\partial t}(0, x, y) = u_1(x, y) \\ u(t, x, y) = 0, \quad \forall (x, y) \in \partial\Omega \backslash M \end{cases} \tag{3.17}$$

where $\gamma, \lambda, \beta \in (0, 1]$ and the edge-stopping function is:

$$\begin{aligned} &\xi_u : [0, \infty) \to [0, \infty) \\ &\xi_u(s) = \varepsilon \left(\frac{K(u)}{\alpha \log_{10}(s + K(u))^k + \nu} \right)^{1/3} \end{aligned} \tag{3.18}$$

with $\alpha, \varepsilon \in (0, 1), \nu \in (1, 5], k \in \{1, 2, 3, 4\}$ and $K(u(x, y, t)) = |\eta\, median(\|\nabla u\|) - \delta t|, \eta \in (1, 3)$ and $\delta \in (0, 1)$. The next numerical approximation algorithm is built for it:

$$\begin{aligned} u_{i,j}^{n+1} = u_{i,j}^{n} \left(\frac{4\lambda - 2\beta(1 - 1_M)}{2\gamma + \lambda^2} \right) + u_{i,j}^{n-1} \frac{\lambda^2 - 2\gamma}{2\gamma + \lambda^2} \\ - \varsigma \sum_{q \in N_p} \xi_u\left(\left\|\nabla^{p,q} u\right\|\right) \cdot \nabla^{p,q} u + u_{i,j}^0 \beta(1 - 1_M) \end{aligned} \tag{3.19}$$

This completion method inpaints successfully the images, in both clean and noisy conditions. It removes the additive noise and unintended effects, and pre-serves and sharpens the features, given its hyperbolic character. It provides better results than some existing PDE inpainting models. See the averaged PSNR values achieved by our hyperbolic PDE-based technique and other approaches, in Table 3.4. A structural inpainting example is described in Fig. 3.3 [19].

A non-variational second-order PDE-based inpainting technique that was disseminated in [20] is based on the following nonlinear anisotropic diffusion model:

$$\begin{cases} \frac{\partial u}{\partial t} - \psi^u(\|\nabla u\|)\nabla \cdot (\varphi_u(\|\nabla u\|)\nabla u) + \lambda(1 - 1_\Gamma)(u - u_0) = 0, \quad \forall (x, y) \in \Omega \\ u(x, y, 0) = u_0(x, y), \quad \forall (x, y) \in \Omega \\ u(t, x, y) = 0, \quad \forall (x, y) \in \partial\Omega \end{cases} \tag{3.20}$$

Table 3.4 PSNR values achieved by several inpainting techniques

Interpolation method	Average PSNR (dB)
This hyperbolic PDE model	31.27
TV Inpainting	29.86
Harmonic Inpainting	28.42
Navier–Stokes based inpainting	30.32

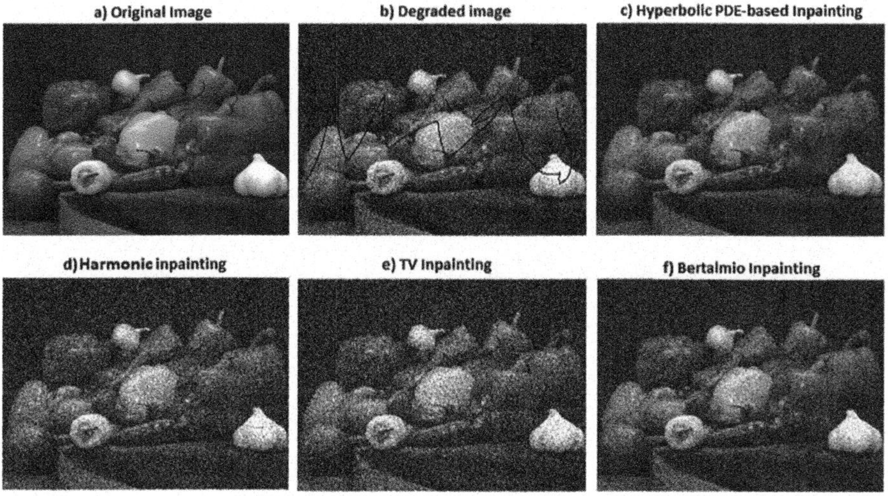

Fig. 3.3 Inpainting results of our method and other approaches

where $\lambda \in (0, 1]$, $\Gamma \subset \Omega \subseteq R^2$, the diffusivity function is

$$\varphi_u : [0, \infty) \to [0, \infty)$$
$$\varphi_u(s) = \delta\left(\frac{\eta(u)}{\xi s^k + \nu \log_{10}(\eta(u))}\right)^{1/k+1} \tag{3.21}$$

where $\delta \in (0, 2)$, $\xi \in (1, 5]$, $\nu \in (0, 1)$, $k \in \{1, 2, 3, 4\}$, $\eta(u(x, y, t)) = |\varepsilon\mu(\|\nabla u\|) + \zeta t|$, $\varepsilon > 1$, $\zeta \in (0, 1)$ and the second function is

$$\psi : [0, \infty) \to [0, \infty)$$
$$\psi(s) = \gamma(\alpha s^\tau + \beta)^{1/\tau+1} \tag{3.22}$$

with $\alpha, \gamma \in (0, 3]$, $\beta \in (0.3, 5]$ and $r \in (0, 2]$.

The well-posedness of this PDE model was demonstrated rigorously in [20]. The next iterative numerical approximation scheme was also proposed in [20]:

$$u_{i,j}^{n+1} = u_{i,j}^n\Big(1 - \lambda(1 - 1_\Gamma) + \psi_{i,j}\Big(\varphi_{i+\frac{1}{2},j} + \varphi_{i-\frac{1}{2},j} + \varphi_{i,j+\frac{1}{2}}\varphi_{i,j-\frac{1}{2}}\Big)\Big)$$

$$+ u^n_{i+1,j}\psi_{i,j}\varphi_{i+\frac{1}{2},j} + u^n_{i-1,j}\psi_{i,j}\varphi_{i-\frac{1}{2},j} + u^n_{i,j+1}\psi_{i,j}\varphi_{i,j+\frac{1}{2}}$$
$$+ u^n_{i,j-1}\psi_{i,j}\varphi_{i,j-\frac{1}{2}} + u^0_{i,j}\lambda(1 - 1_\Gamma) \tag{3.23}$$

where $\varphi_{i\pm\frac{1}{2},j} = \frac{\varphi_{i\pm1,j}+\varphi_{i,j}}{2}$, $\varphi_{i,j\pm\frac{1}{2}} = \frac{\varphi_{i,j\pm1}+\varphi_{i,j}}{2}$ and $u^n_{0,j} = u^n_{1,j}, u^n_{I,j} = u^n_{I+1,j}, u^n_{i,0} = u^n_{i,1}, u^n_{i,J} = u^n_{i,J+1}$.

Method comparison results are provided in the next table and figure. The average PSNR and MSE values achieved by several methods are displayed in Table 3.5 and the results generated by them on a given image are displayed in Fig. 3.4.

The well-known inpainting models based on PDEs of higher order usually have a non-variational character. That is the case with the third order PDE-based inpainting techniques that are based on the Navier–Stokes equations. Such an influential model

Table 3.5 Method comparison: average PSNR and MSE values

Inpainting technique	Average PSNR (dB)	Average MSE
The proposed AD Inpainting	36.1338	15.8380
Harmonic Inpainting	29.6887	69.8576
Total Variation Inpainting	34.2561	24.4047
TV-H^{-1} Inpainting	36.3805	14.9634

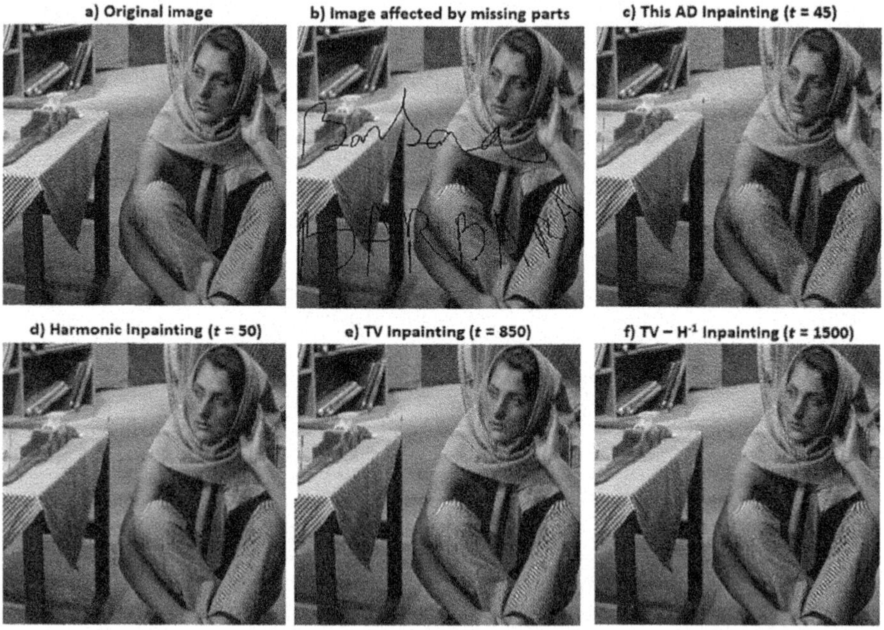

Fig. 3.4 *Barbara* image inpainted by several approaches

was introduced by Bertalmio et al. in 2000 [21]. Their inpainting algorithm propagates the required information in the direction of the image's isophotes [21], being characterized by the equation:

$$\frac{\partial u}{\partial t} = \nabla^{\perp} u \cdot \nabla \Delta u \tag{3.24}$$

where $\nabla^{\perp} u$ represents the perpendicular gradient of image u. This constitutes a transport equation for image smoothness which is modeled by the image's Laplacian along the level lines of the image. An additional anisotropic diffusion is introduced to avoid the level-line crossing, leading to the following nonlinear PDE:

$$\frac{\partial u}{\partial t} = \nabla^{\perp} u \cdot \nabla \Delta u + \nabla \cdot (g(|\nabla u|)\nabla u) \tag{3.25}$$

with $\nu > 0$. The purpose is to evolve this equation to a steady-state solution, where $\nabla^{\perp} u \cdot \nabla \Delta u = 0$, that ensures the information is constant in the isophotes' direction.

The relation between this inpainting scheme and the Navier–Stokes equations for fluid dynamics was demonstrated in [22]. The research performed in the Navier–Stokes equation-based inpainting domain is mostly inspired by this interpolation framework [23]. Also this third-order PDE-based model illustrates that both the gradient direction and gray-scale values must be propagated inside the inpainting area, and shows why the high-order PDEs are required for an effective completion. This diffusion-based inpainting solution restores successfully the images deteriorated by missing regions, being very useful for undesired text and object removal. An old photo inpainting using this model is displayed in Fig. 3.5 [21].

Curvature-driven Diffusion (CDD) Inpainting represents an well-known third-order PDE-based structural interpolation approach developed by Chan and Shen [1, 15, 24]. CDD Inpainting was elaborated as a solution to fix the drawbacks of their previous TV Inpainting model. The CDD-based reconstruction uses the curvature information of the level lines, being characterized by the next form:

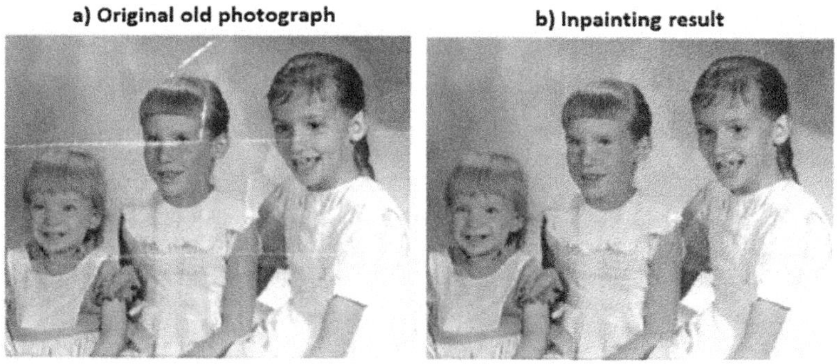

Fig. 3.5 Old photo recovered by the inpainting technique of Bertalmio et al.

$$\begin{cases} \frac{\partial u}{\partial t} = \nabla \cdot \left(\frac{g(\kappa)}{|\nabla u|} \nabla u \right) - \lambda_D (u - u_0) \\ u(0, x, y) = u_0 \end{cases} \tag{3.26}$$

where the isophote curvature $\kappa = \nabla \cdot \left[\frac{\nabla u}{|[\nabla u]|} \right]$ and the positive and continuous function $g(s) = s^p, p \geq 1$ [15, 24]. CDD Eq. (3.26) diffuses the smoothness perpendicularly to the level lines, while preserving their direction and being able to connect them across large distances. That makes the CDD Inpainting model a better interpolation solution than TV Inpainting and other second-order PDE-based methods. Anyway, CDD Inpainting model has its own drawback, such as its noise sensitive character and its rather poor performance in noisy conditions [24].

A finite difference-based numerical solver for (3.26) is provided in [24]. The flux of this curvature-driven diffusion model is as $j = -\frac{g(\kappa)}{|\nabla u|} \nabla u$. So, the CDD equation is re-written as $\frac{\partial u}{\partial t} = -\nabla \cdot j$ and discretized using the numerical scheme:

$$u^{n+1} = u^n - \Delta t \left[\nabla \cdot j \right]^{(n)} \tag{3.27}$$

where $t = n\Delta t$ and $\left[\nabla \cdot j \right]^{(n)}$ is the discretization of the divergence $\nabla \cdot j$, which is achieved using the half-point central differences for divergence operator [24]. The iterative algorithm (3.27) converges stable to the reconstruction output. CDD Inpainting performs effectively a variety of completion tasks, such as disocclusions, reconstruction of old photos damaged by scratches, text and object removals. An example of CDD Inpainting-based text removal is displayed in Fig. 3.6 [15].

The fourth-order PDE-based inpainting models represent another important category of completion methods using high-order PDEs. One of them is the influential Cahn–Hilliard Inpainting framework [1, 15, 25]. The Cahn–Hilliard equation is a nonlinear fourth-order diffusion equation that originates in the material science, but the well-known inpainting model uses on a modified variant of this PDE, which is based on an added fidelity term. It has the following form [1, 25]:

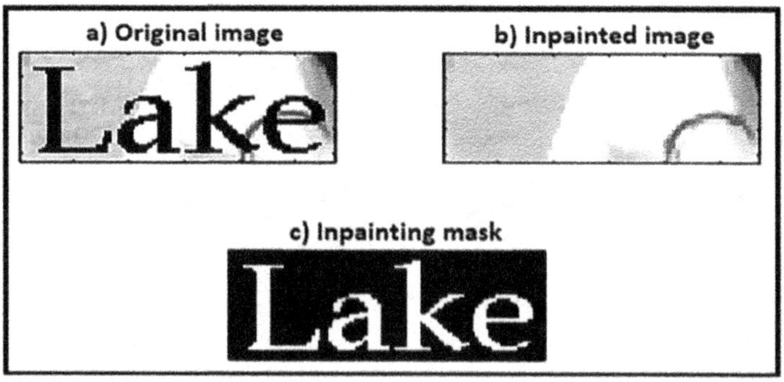

Fig. 3.6 Text removal obtained with CDD Inpainting

$$\begin{cases} u_t = -\Delta\left(\varepsilon \Delta u - \frac{1}{\varepsilon} W'(u)\right) - \lambda(x, y)(u - u_0), & in\ \Omega \\ \frac{\partial u}{\partial \nu} = \frac{\partial \Delta u}{\partial \nu} = 0, & on\ \partial\Omega \end{cases} \tag{3.28}$$

where deteriorated observation $u_0 \in L^2(\Omega)$, the nonlinear double-well potential function $W(u) = u^2(u-1)^2$, $\lambda(x, y) = \begin{cases} \lambda_0 \geq 1, & (x, y) \in \Omega \backslash D \\ 0, & (x, y) \in D \end{cases}$ and D is the inpainting area.

The nonlinear fourth-order PDE-based model (3.28) is well-posed, its validity being demonstrated in [25]. The solution of the Cahn–Hilliard Inpainting model is computed numerically applying a consistent, fast-converging and stable numerical approximation algorithm that is the *convexity splitting* fast solver [25]. Cahn–Hilliard Inpainting provides effective interpolation results and outperforms nonlinear second-order PDE-based models. It generates a smooth continuation of level lines into the inpainting domain *D*, like the curvature-based schemes, but converges faster than CDD Inpainting [24]. A completion example using Cahn–Hilliard Inpainting is displayed in Fig. 3.7. It describes an inpainting result after 1000 iterations [15].

TV-H^{-1} Inpainting model represents another important fourth-order PDE-based inpainting approach [15, 26]. It is described by the next equation:

$$u_t = \Delta p + \lambda(x, y)(u_0 - u), \quad p \in \partial TV(u) \tag{3.29}$$

where $\lambda(x, y)$ is defined as in the previous case and $\partial TV(u)$ is the subdifferential of

$$TV(u) = \begin{cases} |Du|(\Omega), & if\ |u| \leq 1\ a.e.\ in\ \Omega \\ +\infty, & otherwise \end{cases} \tag{3.30}$$

with $|Du|(\Omega)$ representing the total variation of *u*. It was proved that stationary equation $\Delta p + \lambda(u_0 - u) = 0$ admits a solution $u \in BV(\Omega)$ [26], which means Eq. (3.29) has a steady state. The TV-H^{-1} Inpainting is solved applying the next

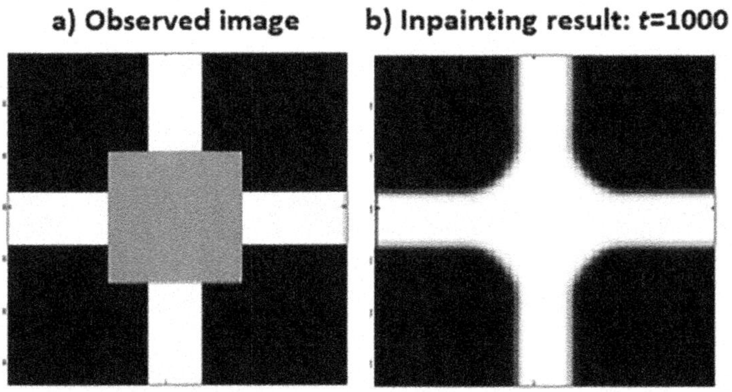

Fig. 3.7 Cahn–Hilliard Inpainting-based reconstruction example

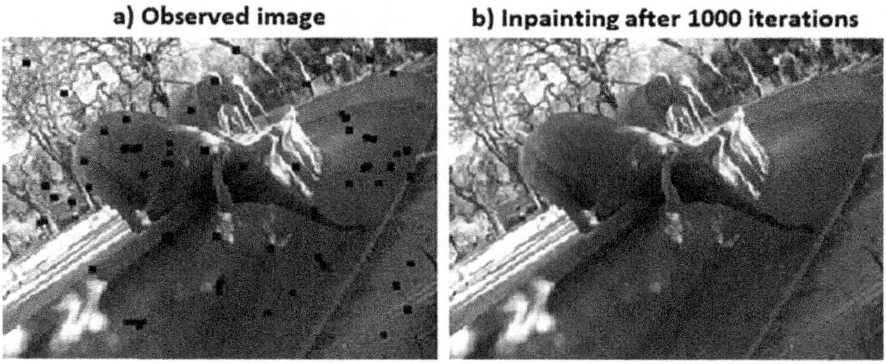

Fig. 3.8 TV-H^{-1} Inpainting example: $\lambda_0 = 10^2$, $\epsilon = 0.01$

time-stepping numerical approximation algorithm:

$$\frac{U_{k+1} - U_k}{\Delta t} + C_1 \Delta\Delta U_{k+1} + C_2 U_{k+1} = C_1 \Delta\Delta U_k - \Delta\left(\nabla \cdot \left(\frac{\nabla U_k}{|\nabla U_k|_\epsilon}\right)\right) + C_2 U_k + \lambda(u_0 - U_k) \tag{3.31}$$

where $C_1 > \frac{1}{\epsilon}$ and $C_2 > \lambda_0$.

TV-H^{-1} Inpainting model reconstructs properly the image and outperforms second-order TV-based inpainting schemes, but also requires a lot of iterations, due to the complexity of its numerical solver. This is illustrated by the example in Fig. 3.8, where the observation u_0 in (a) is inpainted successfully in $t = 1000$ iterations, with parameters $\lambda_0 = 10^2$, $\epsilon = 0.01$, the interpolation output being displayed in (b) [26].

Another well-known fourth-order PDE-based inpainting algorithm is *low curvature image simplifiers* (LCIS)-based inpainting [27], characterized by equation:

$$u_t = -\nabla \cdot (g(|\nabla u|)\nabla \Delta u) + \lambda(u_0 - u) \tag{3.32}$$

where $g(s) = \frac{1}{1+s^2}$. The nonlinear PDE in (3.32) leads to

$$u_t = -\nabla \cdot (\arctan(\Delta u)) + \lambda(u_0 - u) \tag{3.33}$$

that represents a gradient flow in L^2 for the next energy functional:

$$E(u) = \int_\Omega G(\Delta u)dxdy + \frac{1}{2}\int_\Omega \lambda(u_0 - u)^2 \tag{3.34}$$

with $G'(s) = \arctan(s)$. According to the convexity splitting method, it is written as $E(u) = E_1(u) - E_2(u)$, where

$$\begin{cases} E_1(u) = \int\limits_{\Omega} \frac{C_1}{2}(\Delta u)^2 dxdy + \frac{1}{2}\int\limits_{\Omega} \frac{C_2}{2}|u|^2 dxdy \\ E_2(u) = \int\limits_{\Omega} -G(\Delta u) + \frac{C_1}{2}(\Delta u)^2 dxdy + \frac{1}{2}\int\limits_{\Omega} -\lambda(u_0 - u)^2 + \frac{C_2}{2}|u|^2 dxdy \end{cases} \quad (3.35)$$

and C_1 and C_2 are selected so that E_1 and E_2 are strictly convex. One obtains the next time-stepping scheme that solves numerically the PDE-based model:

$$\frac{U_{k+1} - U_k}{\Delta t} + C_1\Delta^2 U_{k+1} + C_2 U_{k+1} = -\Delta(\arctan \Delta U_k) + C_1\Delta^2 U_k + \lambda(u_0 - U_k) + C_2 U_k \quad (3.36)$$

LCIS Inpainting provides effective inpainting results too, but given the computational expensive numerical approximation algorithm (3.36), it does not operate fast.

3.2 Nonlinear PDE-Based Image Compression Techniques

The image compression represents an important sub-domain of both the image processing and data compression fields. The purpose of compression is reducing the image file size without losing too much information and maintaining a high level of visual quality, in order to facilitate the storage and transmission processes. A compression operation encodes the image information using fewer bits than its original representation and may have an either lossy or lossless character [28].

The lossless techniques recover perfectly the compressed image in the decompression stage, no information being lost. The lossless compression algorithms include Huffman encoding, Run Length Encoding (RLE), Arithmetic coding, LZW encoding and Area coding [28–30]. Digital formats like GIF, BMP and PNG are based on lossless coding.

The lossy coding approaches provide much higher compression ratios than lossless methods, but the decompressed images generated by them are not identical to the original versions. They include the vector quantization, transformation coding, fractal coding, Block Truncation Coding or sub-band coding [29]. Some popular lossy compression-based image formats include JPEG, JPEG 2008 and other JPEG versions, and MPEG with its variants (for videos) [28–30].

An important category of lossy image compression schemes contains nonlinear partial differential equation-based methods. The PDE-based image compression domain represents an obvious application area of PDE-based inpainting, since the PDE-based interpolation models are used in the decompression stage.

In the image compression process stage one selects a set of significant pixels that can be coded efficiently and stored according to an algorithm. A sparsification is performed on that image, to reduce the data to a well-adapted set of significant sparse points. Various sparsification and coding models have been developed. The

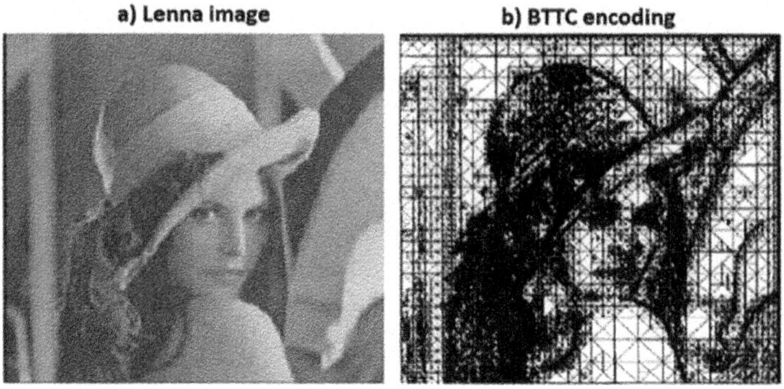

Fig. 3.9 BTTC triangulation of *Lenna* image

random sparsification selects the scattered data points randomly, then encode them. This approach generates low-quality decompressions and high loss of information.

The *B-Tree Triangular Coding* (*BTTC*), introduced by R. Distasi et al., creates an useful sparse point representation and efficient coding of the sparsified data [31]. This encoding model is based on the recursive decomposition of image domain into right-angled triangles positioned in a binary tree. BTTC is a fast coding algorithm, characterized by a convenient computational cost $O(n\log\ n)$ for a number of pixels n, easy to implement and highly parallel. A BTTC example is described in Fig. 3.9. The compression approach in [31, 32] uses a linear interpolation of all these triangles in the decoding stage. This binary tree-based coding works successfully with PDE-based inpainting schemes that interpolate properly its scattered data.

Other tree-based sparsification and coding techniques include the *image coding method using adaptive Delaunay triangulation*, elaborated by Demaret et al. [33], the *tree-based rectangular subdivision scheme* that partitions the original image into rectangular subimages [32] and the *stochastic tree-based image sparsification* [32]. The *edge-based image coding* models has been also used successfully with PDE-based inpainting. They detect the image boundaries, then, only the pixels located on them or in their vicinity are considered as sparse points. The selected pixels are coded applying various encoding algorithms [34]. Clustering-based quantization methods for PDE-based image compression have been also introduced [35]. These quantization-based encoding schemes reduce the number of colors without altering the image quality. Clustering algorithms like K-means can be used for this task. The key pixels corresponding to the detected colors are selected, coded and stored [35].

The scattered data achieved and encoded using the sparsification and coding approaches is decoded and interpolated in the decompression stage using PDE-based inpainting models [1, 36]. The PDE-based compression schemes achieve high values for the compression rate, compression ratio, fidelity and quality. The general form of a PDE-based inpainting model used for decompression is:

$$\frac{\partial u}{\partial t} = (1 - \lambda_D)L(u) - \lambda_D(u - u_0) \tag{3.37}$$

where $L(u)$ is a differential operator determining various compression techniques.

A linear homogeneous diffusion-based image compression approach uses the Harmonic Inpainting, its differential operator having the form:

$$L(u) = -\nabla^2 u = -\Delta u \tag{3.38}$$

The sparsified image is interpolated successfully in decompression step. This compression scheme provides good results for cartoon and depth map images [34, 37]. A linear homogeneous diffusion-based compression method was introduced by Mainberger and Weickert and uses an edge-based encoding algorithm [34]. A boundary detection based on the Marr–Hildreth operator [38] is performed first. The pixels from edges' neighborhoods are encoded applying quantization, subsampling and PAQ coding algorithms and stored with JBIG standard [34]. The decompression is performed applying the linear homogeneous diffusion inpainting. It outperforms JPEG and JPEG 2000 compression standards when applied on cartoon images and measured at the same compression rate. An image compression example using edge-based coding and harmonic interpolation is displayed in Fig. 3.10 [34].

Extended versions of this compression model use differential operators of higher orders. The compression based on the *biharmonic smoothing operator* [39] has the form $L(u) = -\nabla^4 u = -\Delta^2 u$ and provides satisfactory decompression results when used with an effective image coding scheme, like B-Tree Triangular Coding. It provides weaker results when applied along random sparsification. Another derived model is based on *triharmonic smoothing operator* that is a higher-order differential of the form $L(u) := \Delta^3 u$. Triharmonic Inpainting also generates effective image decompression results when applied on the sparse pixels encoded using BTTC [40].

The *absolute monotone Lipschitz extension (AMLE)* model is a second-order PDE-based interpolation scheme applied successfully in image compression [1, 40]. It is based on the second order directional derivative in the gradient direction, being characterized by $L(u) = \partial_{\eta\eta} u$, with $\eta = \frac{\nabla u}{|\nabla u|}$. The AMLE inpainting approach achieves an optimal decompression when applied on a properly selected scattered

Fig. 3.10 Edge-based compression example

pixel set, obtained by BTTC, for example. It has a poorer performance when applied on a randomly selected set of sparse pixels.

The nonlinear second-order anisotropic diffusion-based image compression techniques are characterized by differential operators of the form:

$$L(u) = div(g\left(\|\nabla u\|^2 \nabla u\right) \tag{3.39}$$

where g are the positive and monotonically decreasing diffusivity functions in Sect. 2.1.1.

Some more performant PDE-based image compression techniques are those based on *edge-enhancing diffusion* (*EED*), which preserves and enhances the edges. An effective EED-based encoder was introduced by Galic et al. [41]. That BTTC-EED scheme applies an improved BTTC for image compression and EED for decompression. So, an adaptive BTTC-based sparsification is applied to obtain the scattered interpolation pixels. The image is decomposed into isosceles triangular regions so that within each region it can be recovered by interpolation from vertices. This triangle-based decomposition is stored into a binary tree structure [41]. That tree is stored by traversing it (pre-order or level-order) and storing it one bit per node: 1 for a node that has children, and 0 for a leaf. A vertex mask is obtained. To code grey values in all vertices, one first zig-zag traverses the sparse image created with the binary tree structure and store it in a sequence encoded with Huffman algorithm. The coding is next enhanced by a final quantization [41].

The decompression process consists of decoding and interpolation. The vertex mask is recovered from the encoding and so results the sparse image. Next, the scattered data interpolation is performed using the EED model [41]. This BTTC + EED approach replaces the linear interpolation of the BTTC-L model [31] to an edge enhancing diffusion-based inpainting applied within each triangle of the sparsified image [42]. The EED-based interpolation is based on the diffusion operator:

$$L(u) := \nabla \cdot \left(g\left(\nabla u_\sigma u_\sigma^{\mathrm{T}}\right) \nabla u\right) \tag{3.40}$$

where diffusivity function g is positive, monotonic decreasing and converges to 0.

BTTC-EED scheme produces an effective sparse image interpolation and outperforms other PDE-based inpainting models in the decompression step, providing better *average absolute error* (AAE) and mean squared error scores, displayed in Table 3.6. It outperforms also JPEG standard at the same compression rate, but it is outperformed by JPEG 2000 coding [41, 42]. Some BTTC-EED results (sparsifications and decompressions) at several compression rates are displayed in Fig. 3.11.

Some improved versions of this compression method were also introduced. One of them is the Q64+BTTC(L)-EED image codec introduced by Galic et al. [41]. A more effective EED-based compression technique is *Rectangular subdivision with edge-enhancing diffusion* (*R-EED*) codec introduced by Schmaltz et al. [43]. Their technique performs a rectangular subdivision of the image in compression stage. In this approach a line is saved using 3 points. Each time the quality of the image reconstruction is checked: if MSE value exceeds a given threshold, then the image is

Table 3.6 Method comparison: AAE and MSE values

PDE-based compression method	AAE	MSE
Harmonic smoothing	16.98	611.5
Biharmonic smoothing	15.79	615.5
Triharmonic smoothing	18.69	807.9
AMLE model	17.33	631.7
Charbonnier diffusion	21.80	987.0
BTTC-EED technique	14.58	591.7

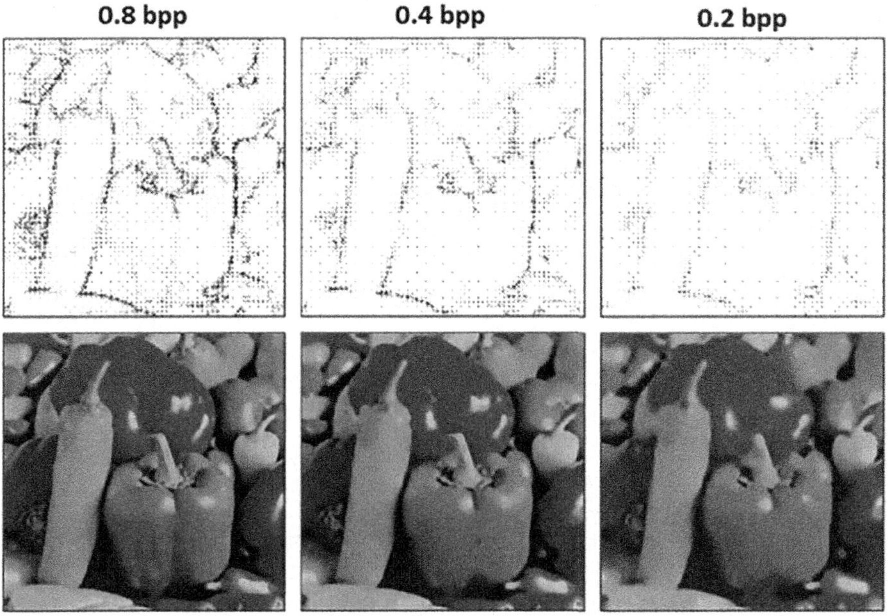

Fig. 3.11 BTTC-EED image compression results at several rates

split into 2 sub-images saving a line between them. The sub-images are then saved in a recursive manner. The saved pixels are quantized and coded applying a entropy coding algorithm [43]. A rectangular subdivision example is described in Fig. 3.12.

The R-EED decompression is performed applying an EED-based inpainting model. The operator $L(u) = div\left(g\left(\nabla u_\sigma u_\sigma^T\right)\nabla u\right)$ is used with Charbonnier diffusivity function in Sect. 2.1.1. An optimization is performed on this function, by selecting an optimal k parameter that depends on the image and compression ratio [43]. R-EED based compression outperforms BTTC-EED algorithm and all other PDE-based compression schemes, and also JPEG. It also provides a better decompression than JPEG 2000, when compared at comparable high compression ratios. While R-EED works better for grayscale images, it is outperformed by JPEG 2000

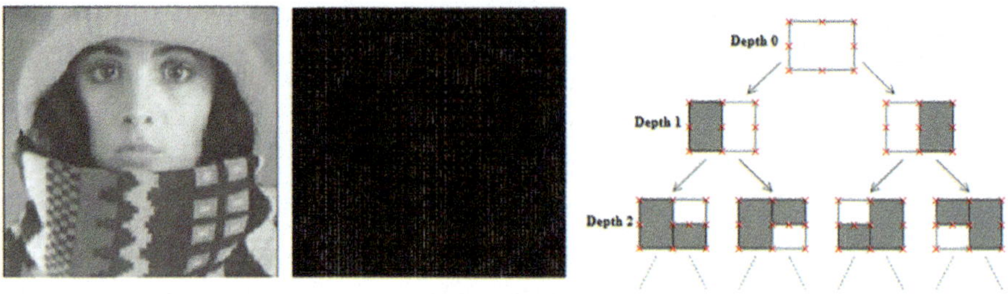

Fig. 3.12 Rectangular subdivision process example

for color images. Some method comparison results at various ratios are described in Fig. 3.13.

An improved R-EED color image compression method was proposed by Peter and Weickert [44]. It provides more attention to encoding of the luma channel of the YCbCr colorspace and outperforms JPEG 2000 for high compression rates.

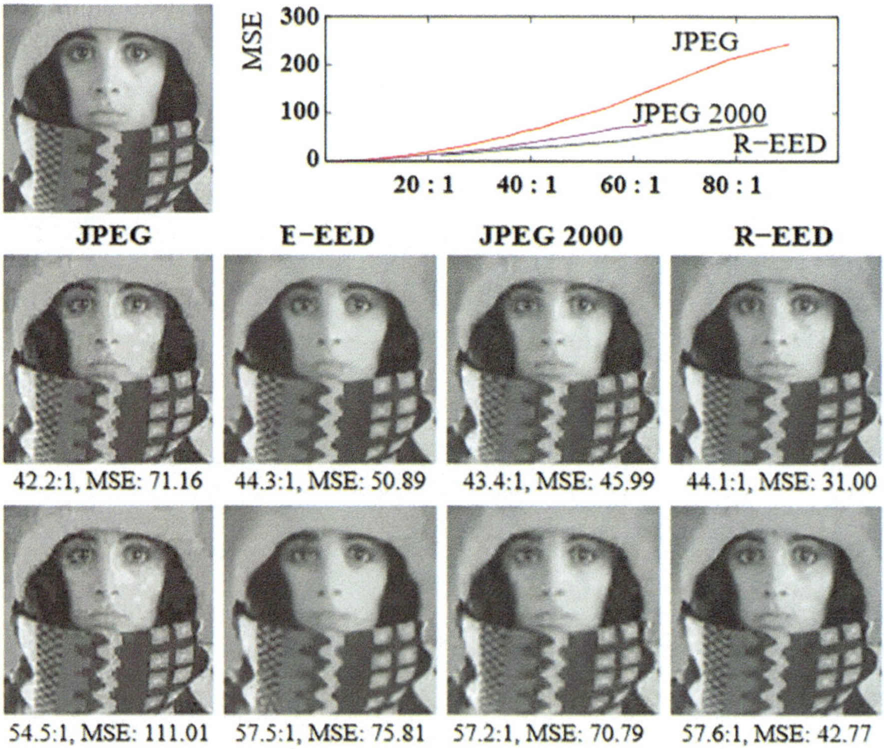

Fig. 3.13 Compression results of various techniques and method comparison

Our own contributions in the nonlinear PDE-based image compression field were disseminated in [45, 46]. Thus, we developed a PDE-based compression technique using edge information [45]. Unlike other PDE-based techniques, it uses nonlinear diffusion schemes for both the image compression and decompression tasks. First, it detects the edges using the second-order anisotropic diffusion-based model:

$$\begin{cases} \frac{\partial u}{\partial t} - \delta\big(\nabla\big(K_{x,y} * u\big)\big)\nabla \cdot (\psi(\nabla u)\nabla u) + \alpha(u - u_0) = 0 \\ u(x, y, 0) = u_0(x, y), \quad \forall(x, y) \in \Omega \\ u(t, x, y) = 0, \quad \forall(x, y) \in \partial\Omega \end{cases} \tag{3.41}$$

where $\alpha \in [0, 1)$, $u_0 = Im * K_{x,y}$, Im is the original image, $K_{x,y}$ is a 2D filter kernel,

$$\psi : [0, \infty) \to [0, \infty), \psi(s) = \varepsilon\left(\frac{\xi(u)}{\left|\beta s^2 + \zeta \ln \xi(u)\right|}\right)^{\frac{1}{3}} \tag{3.42}$$

where $\beta \in (2, 4]$, $\varepsilon, \zeta \in (0, 1)$, $\xi(u) = \nu|\mu(\nabla u) + median(\nabla u)|$, $\nu \in [1, 3)$. Also, $\delta(s) = \kappa \sqrt[r+1]{\upsilon s^r + \rho}$, with $\kappa, r \in (0, 1)$, $\upsilon, \rho \in [1, 2)$. The next explicit finite difference-based numerical approximation scheme is obtained [45]:

$$\begin{aligned} u_{i,j}^{n+1} &= u_{i,j}^n(1 - \alpha) + u_{i,j}^0\alpha + \delta_{i,j}\Big(\psi_{i+\frac{1}{2},j}\Big(u_{i+1,j}^n - u_{i,j}^n\Big) - \psi_{i-\frac{h}{2},j}\Big(u_{i,j}^n - u_{i-1,j}^n\Big) \\ &\quad + \psi_{i,j+\frac{1}{2}}\Big(u_{i,j+1}^n - u_{i,j}^n\Big) - \psi_{i,j-\frac{1}{2}}\Big(u_{i,j}^n - u_{i,j-1}^n\Big)\Big) \end{aligned} \tag{3.43}$$

where $\psi_{i\pm\frac{h}{2},j} = \frac{\psi_{i\pm h,j}+\psi_{i,j}}{2}$, $\psi_{i,j\pm\frac{h}{2}} = \frac{\psi_{i,j\pm h}+\psi_{i,j}}{2}$. One obtains a binary image determining the absolute difference between 2 states and applying a thresholding procedure [45]:

$$u_{i,j}^b = \begin{cases} 1, \ if \ \left|u_{i,j}^{m+T} - u_{i,j}^m\right| \geq w\frac{\xi(u^{m+T})+\xi(u^m)}{2} \\ 0, \ otherwise \end{cases},$$
$$\forall i \in \{1, \ldots, I\}, \quad \forall j \in \{1, \ldots, J\}, \quad w \in (0, 3) \tag{3.44}$$

Some morphological operations, such as dilation, thinning, gap filing and small spot removal, are applied to that binary image. The result represents the edge detection and the pixels located in the vicinity of the extracted boundaries are next coded. A lossless RLE-inspired coding scheme is applied to the sparse pixels. The sparse image is transformed into row vector R and at each step one determines the 3-uple $[R(i), n_i, z_i]$, where $R(i) = \cdots = R(i + n_i - 1)$ and z_i is the number of consecutive 0s after last occurrence. The current 3-uple is appended to coding vector: $C := [C, R(i), n_i, z_i]$.

In the decompression stage, a decoding is applied to the encoded pixels, first. Next, a completion is performed on the sparsified image, applying the next nonlinear fourth-order PDE-based structural inpainting model [45]:

$$\begin{cases} \frac{\partial u}{\partial t} + \lambda\delta(\nabla u)\nabla^2\big(\varphi(\Delta u)\nabla^2 u\big) - \eta\, div(\psi(\nabla u)\nabla u) + (1 - 1_\Gamma)(u - u_0) = 0 \\ u(x, y, 0) = u_0(x, y), \quad \forall(x, y) \in \Omega \\ \frac{\partial u}{\partial \vec{n}} = 0 \\ u(t, x, y) = 0, \quad \forall(x, y) \in \partial\Omega \end{cases} \tag{3.45}$$

where $\lambda \in [1.4, 2)$, $\eta \in (0.5, 1]$, $\varphi : [0, \infty) \to [0, \infty)$, $\varphi(s) = \gamma\sqrt[k]{\frac{\xi(u)}{\varsigma|\xi(u)+s|^{k+1}+\tau}}$ and $\tau \in (2, 4]$, $\gamma, \varsigma \in [0.5, 1)$, $k \in \{2, \ldots, 5\}$.

A numerical approximation algorithm is constructed for (3.45) using the same quantization of the space and time coordinates [45]. Since the PDE is re-written as $\frac{\partial u}{\partial t} + (1 - 1_\Gamma)(u - u_0) = \eta\Big(\frac{\partial}{\partial x}(\psi(|\nabla u|)u_x) + \frac{\partial}{\partial y}\big(\psi(|\nabla u|)u_y\big)\Big) - \lambda\delta(\nabla u)\nabla^2\big(\varphi(\Delta u)\nabla^2 u\big)$, its left term is approximated as $\frac{u_{i,j}^{n+\Delta t} - u_{i,j}^n}{\Delta t} + (1 - 1_\Gamma)\Big(u_{i,j}^n - u_{i,j}^0\Big) = u_{i,j}^{n+\Delta t}\frac{1}{\Delta t} + u_{i,j}^n\big(1 - 1_\Gamma - \frac{1}{\Delta t}\big) - u_{i,j}^0(1 - 1_\Gamma)$. The second-order diffusion-based component, $\eta\Big(\frac{\partial}{\partial x}(\psi(|\nabla u|)u_x) + \frac{\partial}{\partial y}\big(\psi(|\nabla u|)u_y\big)\Big)$, is discretized as

$$\eta\Big(\psi_{i+\frac{1}{2},j}\Big(u_{i+1,j}^n - u_{i,j}^n\Big) - \psi_{i-\frac{h}{2},j}\Big(u_{i,j}^n - u_{i-1,j}^n\Big) \\ +\psi_{i,j+\frac{1}{2}}\Big(u_{i,j+1}^n - u_{i,j}^n\Big) - \psi_{i,j-\frac{1}{2}}\Big(u_{i,j}^n - u_{i,j-1}^n\Big)\Big).$$

The fourth-order diffusion term, $\lambda\delta(\nabla u)\nabla^2\big(\varphi(\Delta u)\nabla^2 u\big)$, is approximated next as $\lambda\delta_{i,j}\Delta\varphi_{i,j} = \lambda\delta_{i,j}\frac{\varphi_{i+1,j}+\varphi_{i-1,j}+\varphi_{i,j+1}+\varphi_{i,j-1}-4\varphi_{i,j}}{h^2}$, where $\varphi_{i,j} = \varphi\big(\big|\Delta u_{i,j}\big|\big)\nabla^2 u_{i,j}$ and $\Delta u_{i,j} = \nabla^2 u_{i,j} = \frac{u_{i+h,j}^n+u_{i-h,j}^n+u_{i,j+h}^n+u_{i,j-h}^n-4u_{i,j}}{h^2}$. If $h = \Delta t = 1$, one obtains the numerical scheme

$$u_{i,j}^{n+1} = u_{i,j}^n\Big(1_\Gamma - \eta\Big(\psi_{i+\frac{1}{2},j} + \psi_{i-\frac{1}{2},j} + \psi_{i,j+\frac{1}{2}} + \psi_{i,j-\frac{1}{2}}\Big)\Big) \\ + \eta\Big(u_{i+1,j}^n\psi_{i+\frac{1}{2},j} + u_{i-1,j}^n\psi_{i-\frac{1}{2},j} + u_{i,j+1}^n\psi_{i,j+\frac{1}{2}} + u_{i,j-1}^n\psi_{i,j-\frac{1}{2}}\Big) \\ + u_{i,j}^0(1 - 1_\Gamma) - \lambda\delta_{i,j}\big(\phi_{i+1,j} + \phi_{i-1,j} + \phi_{i,j+1} + \phi_{i,j-1} - 4\phi_{i,j}\big) \tag{3.46}$$

for $n = 0, 1, \ldots, N$, that is stable and consistent to the diffusion-based model [45].

This PDE-based compression approach was tested successfully on many image datasets and achieved good scores of compression rate, compression ratio, fidelity and quality. It outperforms BTTC-L and the most PDE-based schemes, is slightly better than JPEG at high compression rates, while it is outperformed by JPEG 2000 and R-EED. The main steps of a compression process are displayed in Fig. 3.14 and some method comparison results, are presented in Table 3.7 [45].

Another nonlinear PDE-based image compression framework developed by us applies a feature keypoint-based sparsification algorithm in the coding stage [46]. Our approach extracts all SIFT, SURF, MSER, BRISK, ORB, FAST, KAZE and Harris keypoints of the analyzed image but uses only the strongest k of each type.

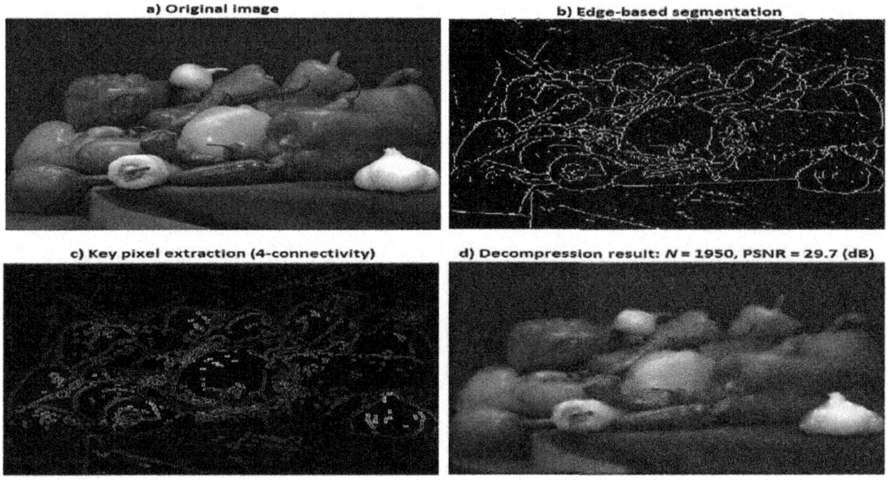

Fig. 3.14 Main stages of a compression process, at ratio = 12:1

Table 3.7 Method comparison: PSNR and SSIM at compression rate 0.4 bpp

Compression technique	PSNR (dB)	SSIM
The proposed AD-based technique	27.8777	0.8231
Linear homogeneous diffusion	20.8971	0.6866
Biharmonic smoothing operator	21.4235	0.7230
BTTC-L compression scheme	22.8565	0.7487
BTTC-EED compression	25.6607	0.7738
R-EED codec	28.9994	0.8586
JPEG	27.7475	0.8143
JPEG 2000	28.4545	0.8467

The pixels related to those keypoints, or those positioned in their neighborhood, are used for sparsification. The sparsified image is converted into a vector V and a code $C(V)$ is constructed for this it. At each iteration, the coding algorithm determines $\left[V(i), n_i, n_i^0\right]$, where $V(i) = \cdots = V(i + n_i - 1)$ and $V(i + n_i) = \cdots = V\left(i + n_i + n_i^0 - 1\right) = 0$, and appends it to the code: $C(V) := \left[C(V), \left(V(i), n_i, n_i^0\right)\right]$. The image width to height ratio is finally added to $C(V)$, which represents the compression result [46].

In the decompression stage one applies a decoding algorithm to the values encoded in $C(V)$, to recover the vector V, then a sparse image is obtained from V. It is then reconstructed by the next nonlinear fourth-order PDE-based inpainting model:

$$
\begin{cases}
\frac{\partial u}{\partial t} - \alpha\psi\left(\left|\nabla^2 u\right|\right)\nabla\cdot\left(\delta\left(\left\|\nabla^4 u_\sigma\right\|\right)\nabla u\right) + (1 - 1_\Gamma)(u - u_0) = 0, & \text{on } (0,T)\times\partial\Omega \\
u(0,x,y) = u_0(x,y), \quad \forall(x,y)\in\Omega \\
u(t,x,y) = 0, \quad \text{on } (0,T)\times\partial\Omega \\
\Delta u(t,x,y) = 0, \quad \text{on } (0,T)\times\partial\Omega
\end{cases}
\tag{3.47}
$$

where $\alpha \in [1,2)$, $\Gamma \subset \Omega \subseteq R^2$ is the inpainting region of the sparse image $u_0 : \Omega\backslash\Gamma \to R$ and the positive and monotonic decreasing diffusivity function is

$$
\delta : [0,\infty) \to [0,\infty), \delta(s) = \gamma\left(\frac{\eta(t,x,y)}{\left|\lambda \ln \eta(t,x,y) + \beta s^k\right|}\right)^{\frac{1}{k+1}}
\tag{3.48}
$$

where $\lambda, \gamma \in (0,1), \beta \in (1,3], k \geq 1$ and the conductance function $\eta(t,x,y)$ depends on the evolving image's coordinates and statistics [46]. The other positive function is $\psi(s) = \frac{(\xi s^r + \nu)^{\frac{1}{r+1}}}{\zeta}$, where $\xi, \nu, \zeta \in [1,3)$ and $r \in (0,1)$ [46]. The fourth-order PDE interpolation model (3.47) reconstructs properly the observed image that is known only in the sparse pixel locations. It fulfills the scattered pixel interpolation directing the diffusion toward the inpainting region only [46].

The mathematical validity of this nonlinear fourth-order diffusion-based model is investigated rigorously in [46]. A mathematical treatment is performed on its well-posedness, to demonstrate the existence and unicity of a weak solution for it. So, one transforms (3.47) into the next parabolic equation of porous media type:

$$
\begin{cases}
v_t - \alpha\psi\left(\left|\nabla^2 u\right|\right)\nabla\cdot\left(\delta(\|G_1 * v\|)\cdot\nabla\left(\Delta^{-1}v\right)\right) + \Delta(1 - 1_\Gamma)\left(\Delta^{-1}v - u_0\right) = 0, & \text{on } (0,T)\times\partial\Omega \\
v(0,x,y) = \Delta u_0(x,y), \quad (x,y)\in\Omega \\
v(t,x,y) = 0, \quad \text{on } (0,T)\times\partial\Omega
\end{cases}
\tag{3.49}
$$

where $G_1 = \nabla^2 G_\sigma$ and $v = \nabla^2 u = \Delta u$ [46].

Since $\nabla\cdot\left(\delta(\|G_1 * v\|)\nabla\left(\Delta^{-1}v\right)\right) = \delta'(\|G_1 * v\|)(\nabla G_1 * v)\cdot\nabla\left(\Delta^{-1}v\right) + \delta(\|G_1 * v\|)v$, (3.49) becomes

$$
\begin{cases}
v_t - \alpha\Delta\phi(x,y,v) + \Delta\left((1 - 1_\Gamma)\left(\Delta^{-1}v - u_0\right)\right) = 0, & \text{on } (0,T)\times\partial\Omega \\
v(0,x,y) = v_0(x,y) = \Delta v_0(x,y), \quad (x,y)\in\Omega \\
v(t,x,y) = 0, \quad (x,y)\in\Omega, \quad t\in(0,T)
\end{cases}
\tag{3.50}
$$

where $\phi(x,y,v) = \psi(|v|)\delta'(\|G_1 * v\|)(\nabla G_1 * v)\cdot\nabla\left(\Delta^{-1}v\right) + \delta(\|G_1 * v\|)v$ [46].

So, instead of (3.47) one studies the existence problem for nonlinear parabolic problem (3.50). The next hypotheses are verified [47, 48]:

i. $\psi : [0,\infty) \to [0,\infty)$ is of class C^1 and $\beta_1 \leq \psi'(r) \leq \beta_2, \forall r \in R^+$ where $\beta_1, \beta_2 > 0$
ii. $\delta \in C^1(R), \delta > \rho > 0$, where ρ is a constant
iii. $(\nabla G_1 * v)\cdot\nabla\left(\Delta^{-1}v\right) \geq 0, \forall v \in L^2(\Omega), v \geq 0$.

We note that (ii)–(iii) imply $\phi(x, y, v) \geq \alpha_1 \psi(v)v, \forall v \geq 0$, where $\alpha_1 > 0$. By weak solution to (3.50) we mean a function $v : [0, T] \times \Omega \to R$ such that

$$v \in L^2(0, T; L^2(\Omega)), \quad v_t \in L^2\left(0, T; \left(L^2\right)^*\right) \tag{3.51}$$

where $\left(L^2\right)^*$ is the dual space of $L^2(\Omega)$ with pivotal space $H^{-1}(\Omega)$ that is the dual space of the Sobolev space $H_0^1(\Omega)$ and

$$\int_\Omega v(t, x, y)(-\Delta)^{-1}\varphi(x, y)dxdy = \alpha \int_\Omega \phi(x, y, v(t, x, y))\varphi(x, y)dxdy$$
$$-\int_\Omega (1 - 1_\Gamma)\left(\Delta^{-1}v(t, x, y) - u_0(x, y)\right)\varphi(x, y)dxdy = 0 \tag{3.52}$$

for all $\varphi \in L^2(\Omega)$ [47, 48]. We note that from (3.51) and (3.52) it follows that $v \in C\left([0, T]; H^{-1}(\Omega)\right)$ and $\phi(\cdot, v) \in L^2\left(0, T; H_0^1(\Omega)\right)$ which implies $v = 0$ on $\partial\Omega$.

Proposition 1 If $v_0 = \Delta u_0 \in L^2(\Omega)$ and $v_0 \geq 0$, then, under the hypotheses (i)–(iii), for T sufficiently small, there exists at least one solution $v \in C\left([0, T]; H^{-1}(\Omega)\right) \cap L^2\left([0, T]; H_0^1(\Omega)\right), v \geq 0$ to Eq. (3.50).

Proof We consider

$$\aleph = \left\{ v \in L^2\left(0, T; H_0^1(\Omega)\right); \int_\Omega v^2(t, x, y)dxdy \leq M, \right.$$
$$\left. \int_0^T \int_\Omega |\nabla v(t, x, y)|^2 dtdcdy \leq M, v \geq 0 \quad on(0, T) \times \Omega \right\}$$

and for each $w \in \aleph$ consider equation

$$\begin{cases} v_t - \alpha\Delta(\psi(v)\varsigma) + \Delta(1 - 1_\Gamma)\left(\Delta^{-1}v - u_0\right) = 0 \\ v(0, x, y) = v_0(x, y) = \Delta u_0(x, y) \\ v(t, x, y) = 0 \quad \text{on } (0, T) \times \partial\Omega \end{cases} \tag{3.53}$$

where $\varsigma = \nabla \cdot \left(\delta(\|G_1 * v\|)\nabla\left(\Delta^{-1}w\right)\right)$. For $\forall w \in \aleph$, the operator $A(t) : L(\Omega) \to \left(L^2\right)^*$ defined by ${}_{(L^2)^*}(A(t)v, \varphi)_{L^2(\Omega)} = \alpha \int_\Omega \psi(v)\varsigma\varphi dxdy, \forall\varphi \in L^2(\Omega)$ is monotone, continuous and coercive, that is ${}_{(L^2)^*}(A(t)v, \varphi)_{L^2(\Omega)} = C_1\|\varphi\|^2_{L^2(\Omega)} - C_2\|\varphi\|^2_{(L^2)^\infty}$. Next, by the Theorem 1.2, Chap. 2 in [48] it follows that (3.53), which is

$$\begin{cases} \frac{\partial v}{\partial t} + A(t)v = 0, \quad t \in (0.T) \\ v(0, x, y) = v_0(x, y) \end{cases} \tag{3.54}$$

has a unique solution $v = F(w)$, $v \in L^2(0, T; H_0^1(\Omega))$, $\frac{\partial v}{\partial t} \in L^2(0, T; (L_2)^*)$. Moreover, if multiply (3.53) by v and integrate it on $(0, t) \times \Omega$ we get the next estimate

$$\frac{d}{dt}\int_\Omega v^2(t,x,y)dxdy + \alpha_1 \int_\Omega |\nabla v(t,x,y)|^2 dxdy \le CM + C\int_\Omega |v(t,x,y)|^2 dxdy \tag{3.55}$$

for each $w \in \aleph$, where $C > 0$ and thus we have

$$\frac{d}{dt}\int_\Omega v^2(t,x,y)dxdy + \alpha_1 \int_0^t \int_\Omega |\nabla v(s,x,y)|^2 dsdxdy \le C(M+1)T + \int_\Omega v_0^2(x,y)dxdy \tag{3.56}$$

So, for T sufficiently small and M suitable chosen, we have $F(w) \in \aleph, \forall w \in \aleph$. Moreover, F is continuous from $L^2(0, T; L^2(\Omega))$ to itself and $F(\aleph)$ is compact in $L^2(0, T; L^2(\Omega))$ because by (3.53) we see that $\| \frac{dv}{dt} \|_{L^2(0,T;H^{-1}(\Omega))} \le C, \forall w \in \aleph$. By the Schauder's fixed point theorem [48], there is $w^* \in \aleph$ such that $F(w) = w^*$. Clearly, $u = w^*$ represents a solution to (3.49) as claimed. From the Proposition 1, it follows an existence result for (3.47) via the transformation $u = \Delta^{-1}v$, so the nonlinear PDE model is well-posed. Its unique weak solution is determined by solving numerically the PDE, using the following consistent and fast-converging iterative finite difference-based approximation scheme that is proposed in [46]:

$$\begin{aligned} u_{i,j}^{n+1} &= u_{i,j}^n(\lambda(1_\Gamma - 1) + 1) + \lambda u_{i,j}^0 \\ &\quad + \alpha\psi_{i,j}\Big[\delta_{i+\frac{1}{2},j}\Big(u_{i+1,j}^n - u_{i,j}^n\Big) - \delta_{i-\frac{1}{2},j}\Big(u_{i,j}^n - u_{i-1,j}^n\Big) \\ &\quad + \delta_{i,j+\frac{1}{2}}\Big(u_{i,j+1}^n - u_{i,j}^n\Big) - \delta_{i,j-\frac{1}{2}}\Big(u_{i,j}^n - u_{i,j-1}^n\Big)\Big] \end{aligned} \tag{3.57}$$

This nonlinear diffusion-based compression technique was been applied successfully on numerous images. It works properly for both clean and noisy images, achieving good compression performance metric values [46]. It outperforms many well-known compression approaches, providing better decompression results than compression schemes using random sparsification, edge-based coding or BTTC with various PDE inpainting models. It also outperforms BTTC-L compression and performs slightly better than BTTC-EED and JPEG at higher compression rates, but it is outperformed by JPEG 2000 and R-EED compression, as shown in Table 3.8. A method comparison example at 0.4 bpp compression rate is displayed in Fig. 3.15.

Table 3.8 Average PSNR values of various methods at 0.4 bpp compression rate

Compression technique	Average PSNR (dB)
The proposed 4th-order PDE-based approach	26.3475
Harmonic Inpainting-based compression	20.8495
Charbonnier diffusion-based model	22.3321
BTTC-L	23.1485
BTTC-EED	26.2378
JPEG	26.1075
JPEG 2000	26.8379
R-EED	27.1465

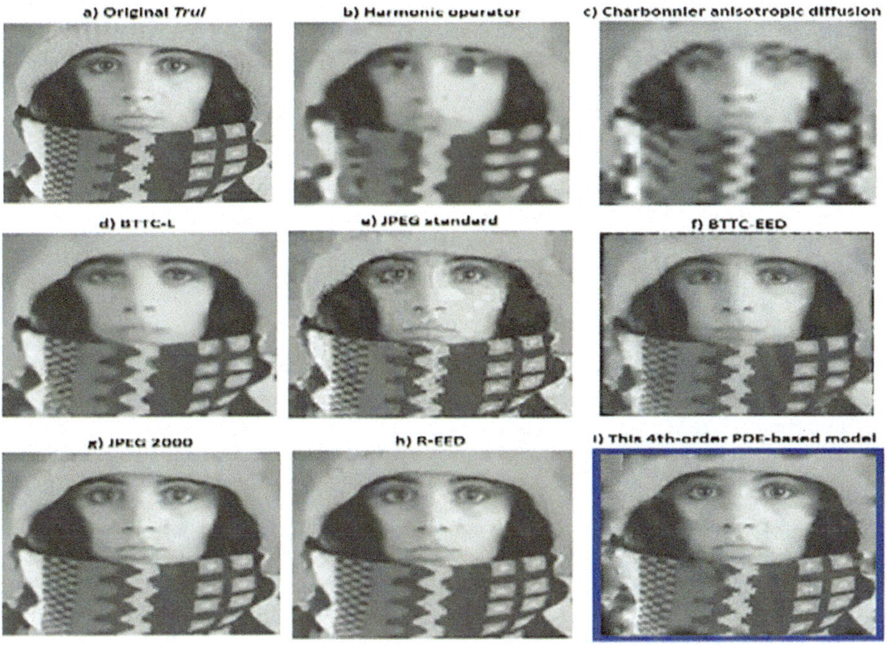

Fig. 3.15 Decompression results of several methods at compression rate of 0.4 bpp

References

1. C. B. Schonlieb, *Partial Differential Equation Methods for Image Inpainting*, Volume 29, Cambridge University Press, 2015.
2. A. A. Efros, T. K. Leung, Texture synthesis by non-parametric sampling, *Proceedings of the International Conference on Computer Vision*, volume 2, pp. 1033-1038, 1999.
3. A. Criminisi, P. Perez, K. Toyama, Region filling and object removal by exemplar-based image inpainting, *IEEE Transactions on Image Processing*, vol. 13, no. 9, pp. 1200–1212, 2004.

4. B. Song, *Topics in Variational PDE Image Segmentation, Inpainting and Denoising*, University of California, 2003.
5. L. Ambrosio, V. M. Tortorelli, Approximation of functionals depending on jumps by elliptic functionals via Γ- convergence, *Comm. Pure Appl. Math.*, 43, pp. 999–1036, 1990.
6. S. Esedoglu, J. Shen, Digital inpainting based on the Mumford-Shah Euler image model, *European Joumal of Applied Mathematics*, 2002.
7. T. Chan, J. Shen, *Morphologically invariant PDE inpaintings*, UCLA CAM Report, pp. 1-15, 2001.
8. P. Getreuer, Total Variation Inpainting using Split Bregman, *Image Processing On Line*, 2, pp. 147–157, 2012.
9. K. Papafitsoros, C. B. Schonlieb, A combined first and second order variational approach for image reconstruction, *J. Math. Imaging Vision*, 48, pp. 308–338, 2014.
10. K. Papafitsoros, C. B. Schoenlieb, B. Sengul, Combined first and second order total variation inpainting using split Bregman, *Image Process. On Line*, vol. 2013, pp. 112–136, Jan. 2013.
11. K. Bredies, M. Holler, A TGV-based framework for variational image decompression, zooming, and reconstruction, Part I: Analytics, *SIAM J. Imag. Sci.*, vol. 8, no. 4, pp. 2814–2850, 2015.
12. M. V. Afonso, J. M. R. Sanches, Blind Inpainting Using l_0 and Total Variation Regularization, *IEEE Transactions on Image Processing*, 24, 7, pp. 2239–2253, 2015.
13. M. Neri, E. R. Zara, Total variation-based image inpainting and denoising using a primal-dual active set method, *Philippine Science Letters*, Vol. 7, No. 1, pp. 97-103, 2014.
14. T. Chan, S. Kang, J. Shen, Euler's elastica and curvature based inpaintings, *SIAM J. Appl. Math.*, 2002.
15. T. Barbu, *Novel Diffusion-Based Models for Image Restoration and Interpolation*, Book Series: Signals and Communication Technology, Springer International Publishing, 2019.
16. T. Barbu, Variational image inpainting technique based on nonlinear second-order diffusions, *Computers & Electrical Engineering*, Volume 54, pp. 345–353, August 2016.
17. T. Barbu, Hybrid Image Interpolation Technique based on Nonlinear Second and Fourth-order Diffusions, *Proc. of the 13th Intl. Symp. on Signals, Circuits and Systems, ISSCS'17*, Iasi, Romania, pp. 1–5, July 13 - 14, 2017.
18. T. Barbu, Nonlinear Anisotropic Diffusion-based Structural Inpainting Framework, *Proceedings of the 13th International Conference TELSIKS'17*, Nis, Serbia, pp. 207–210, 18–20 October 2017, IEEE.
19. T. Barbu, Structural Image Interpolation using a Nonlinear Second-order Hyperbolic PDE-based Model, *Proceedings of the 6th IEEE International Conference on e-Health and Bioengineering, EHB 2017*, Sinaia, Romania, pp. 5–8, 22–24 June 2017.
20. T. Barbu, Second-order Anisotropic Diffusion-based Framework for Structural Inpainting, *Proceedings of the Romanian Academy*, Series A, Volume 19, Issue 2, April–June 2018.
21. M. Bertalmio, G. Sapiro, V. Caselles and C. Ballesters, Image inpainting, *Proc. ACM Conf. Comp. Graphics (SIGGRAPH)*, pp. 417–424, New Orleans, LU, July 2000.
22. M. Bertalmio, A. L. Bertozzi, G. Sapiro, Navier-stokes, fluid dynamics, and image and video inpainting, *Proc. Conf. Comp. Vision Pattern Rec.*, pp. 355–362, Hawai, December 2001.
23. M. A. Ebrahimi, M. Holst, E. Lunasin, The Navier-Stokes-Voight model for image inpainting, *IMA J. Appl. Math.*, 78 (5), pp. 869–894, 2013.
24. T. F. Chan, J. Shen, Non-texture inpainting by curvature-driven diffusions (CDD), *J. Visual Comm. Image Rep.*, 4 (12), pp. 436–449, 2001.
25. M. Burger, L. He, C. Schonlieb, Cahn-Hilliard inpainting and a generalization for grayvalue images, *SIAM J. Imaging Sci.*, 2 (4), pp. 1129–1167, 2009.
26. S. Osher, A. Sole, L. Vese, Image decomposition and restoration using total variation minimization and the H^{-1} norm, *Multiscale Modeling and Simulation: A SIAM Interdisciplinary Journal*, Vol. 1, Nr. 3, pp. 349-370, 2003.
27. C. B. Schoenlieb, A. Bertozzi, Unconditionally stable schemes for higher order inpainting, *Comm. Math. Sci.*, vol. 2, no. 9, pp. 413–457, 2011.
28. R. Gonzalez, R. Woods, *Digital Image Processing*, Prentice Hall, New York, NY, USA, 2nd ed., 2001.

29. V. Bhaskaran, K. Konstantinides, *Image and Video Compression Standards*, Kluwer Academic Press, Boston, 1995.
30. K. Sayood, *Introduction to Data Compression*, Morgan Kaufmann, Third edition, 2005.
31. R. Distasi, M. Nappi, S. Vitulano, *Image compression by B-tree triangular coding*, IEEE Transactions on Communications 45 (9), pp. 1095-1100, 1997.
32. P. T. Peter, *Understanding and advancing PDE-based image compression*, PhD Thesis, Saarland University, 2016.
33. L. Demaret, N. Dyn, A. Iske, *Image compression by linear splines over adaptive triangulations*, Technical report, Dept. of Mathematics, University of Leicester, UK, January 2005.
34. M. Mainberger, J. Weickert, Edge-based image compression with homogeneous diffusion, *Computer Analysis of Images and Patterns*, Lecture Notes in Computer Science, Vol. 5702, Springer, Berlin, pp. 476–483, 2009.
35. L. Hoeltgen, P. Peter, M. Breuß, Clustering-based quantization for PDE-based image compression, *Signal, Image and Video Processing* 12 (3), pp. 411-419, 2018.
36. G. Aubert, P. Kornprobs, *Mathematical problems in image processing: partial differential equations and the calculus of variations*, Vol. 147, Springer Science & Business Media, 2006.
37. J. Chen, F. Ye, J. Di, C. Liu, A. Men, Depth map compression via edge-based inpainting, *Proceedings of the 29th Picture Coding Symposium*, pp. 57–60, Krak6w, Poland, May 2012.
38. D. Marr, E. Hildreth, Theory of Edge Detection, *Proceedings of the Royal Society of London*, Series B, Biological Sciences, 207 (1167), pp. 187–217, 1980.
39. S. Hoffmann, G. Plonka, J. Weickert, Discrete Green's functions for harmonic and biharmonic inpainting with sparse atoms, *International Workshop on Energy Minimization Methods in Computer Vision and Pattern Recognition*, pp. 169–182, 2015.
40. C. Schmaltz, P. Peter, M. Mainberger, F. Ebel, J. Weickert, A. Bruhn, Understanding, optimizing, and extending data compression with anisotropic diffusion, *International Journal of Computer Vision*, 108 (3), pp. 222-240, 2014.
41. I. Galic, J. Weickert, M. Welk, A. Bruhn, A. Belyaev, H.-P. Seidel, Image compression with anisotropic diffusion, *Journal of Mathematical Imaging and Vision*, 31 (2-3), pp. 255-269, 2008.
42. I. Galic, J. Weickert, M. Welk, A. Bruhn, A. Belyaev, H.-P. Seidel, *Towards PDE-based image compression, Variational, Geometric and Level-Set Methods in Computer Vision*, Lecture Notes in Computer Science, vol. 3752, Springer, Berlin, pp. 37–48, 2005.
43. C. Schmaltz, J. Weickert, A. Bruhn, Beating the quality of JPEG 2000 with anisotropic diffusion, *Pattern Recognition*, Lecture Notes in Computer Science, Vol. 5748, Springer, pp. 452–461, 2009.
44. P. Peter, J. Weickert, Colour image compression with anisotropic diffusion, *Proc. of 21st IEEE International Conference on Image Processing*, Paris, France, pp. 4822–4826, 2014.
45. T. Barbu, Segmentation-Based Non-Texture Image Compression Framework Using Anisotropic Diffusion Models, *Proceedings of the Romanian Academy*, Ser. A, Vol. 20, no. 2, pp. 122–130, 2019.
46. T. Barbu, Feature Keypoint-Based Image Compression Technique Using a Well-Posed Nonlinear Fourth-Order PDE-Based Model, *Mathematics*, *8*, 930, MDPI, June 2020.
47. L. Lions, W. Strauss, Some non-linear evolution equations, *Bulletin de la Société Mathématique de France; Bulletin de la S.M.F.*, Paris, France, Volume 93, pp. 43–96, 1965.
48. L. Lions, *Quelques Méthodes de Résolution des Problèmes aux Limites non Linéaires*, Dunod & Gauthier-Villars: Paris, France, 1969.

Chapter 4
Nonlinear Diffusion-Based Multi-scale Image Analysis Methods

The multi-scale image analysis and computer vision techniques based on nonlinear PDE models are discussed in this chapter. The results described here illustrate that many effective multi-scale analysis solutions for various tasks can be achieved by using nonlinear diffusion-based models. Some nonlinear PDE-based scale-space representations are created and used by the high-level feature extraction components of several image analysis techniques proposed by us. Thus, some anisotropic diffusion-based multi-scale texture recognition frameworks are described in the first section. Then, the content-based image recognition using nonlinear PDE-based scale spaces is discussed in the second section. The nonlinear diffusion-based multi-scale image indexing and retrieval solutions are presented in the final section.

4.1 Anisotropic Diffusion-Based Multi-scale Texture Recognition

Multi-scale and multi-resolution analysis approaches, that handle image structures at different scales and resolutions, are applied successfully in various image processing and computer vision domains, including the texture analysis, since the multi-scale and multi-resolution representations allow more flexibility and provide better results than traditional approaches. A scale space representation is created applying a 2D filter kernel to the analysed image at various scales. While many multiscale image analysis techniques use scale spaces based on 2D Gaussian filters and Gaussian derivatives [1], we have introduced more effective scale spaces using the diffusion-based filtering.

Texture analysis represents an important image analysis field that is successfully applied in computer vision areas like object detection, recognition and tracking, image indexing and retrieval, medical imaging and remote sensing. It includes the next subdomains: texture recognition, segmentation, synthesis and retrieval.

T. Barbu, *Digital Image Processing, Analysis and Computer Vision Using Nonlinear Partial Differential Equations*, Studies in Computational Intelligence 1211,
https://doi.org/10.1007/978-3-031-89576-0_4

Texture recognition consists of texture feature extraction and classification. The texture featuring methods are grouped in statistical, structural, model-based and transform-based approaches. Statistical methods include histogram-based schemes [2], moment-based algorithms [3], Gray Level Co-occurrence Matrices (GLCM) [4], Local Binary Patterns (LBP) [5], Binary Gabor Patterns (BGP) [6] and energy variation-based approaches [7]. Structural methods include edge-based techniques [8], morphological operators [9] and SIFT descriptors [10]. The model-based techniques include fractal texture models [11], Markov random field texture models [12] and autoregressive models [13]. Transform-based approaches include feature extraction solutions based on 2D Gabor filters [14], Wavelet transforms [15] and Curvelet transforms [16]. Other featuring methods combine such descriptors [17].

Texture recognition may have either a supervised or an unsupervised character [18]. Supervised machine learning algorithms used for texture classification include K-Nearest Neighbour (K-NN), minimum distance classifier, artificial neural networks (ANN), Support Vector Machines (SVM) and Hidden Markov Models (HMM), while unsupervised classifiers include K-means, hierarchical clustering, Self-organizing Maps (SOM) and Dynamic Time Warping (DTW). Deep learning [19] and graph clustering have been also applied successfully for classification [20].

Many texture recognition methods provide poor results in the presence of noise or when textures have undergone rotations. So, we developed some rotation-invariant texture recognition frameworks using the multi-scale image analysis, that works properly in both normal and noisy conditions. A supervised texture recognition approach was disseminated in [21]. We introduced the next second-order anisotropic diffusion model for creating the scale space:

$$\begin{cases} \frac{\partial u}{\partial t} - \alpha\varphi(|\Delta u_\sigma|)\nabla \cdot (\psi(\|\nabla u_\sigma\|)\nabla u) + \lambda(u - u_0) = 0, & (x,y) \in \Omega \\ u(0,x,y) = u_0(x,y), \quad \forall (x,y) \in \Omega \\ u(t,x,y) = 0, \quad \forall (x,y) \in \partial\Omega; \end{cases} \tag{4.1}$$

where $\alpha \in [1, 2)$, $\lambda \in (0, 1)$ and the considered diffusivity function is:

$$\psi : [0, \infty) \to [0, \infty), \psi(s) = \left(\frac{\eta}{|\delta \log 10(\eta) + \xi s^k|}\right)^{1/k+1} \tag{4.2}$$

where $\delta, \xi \in (0, 1]$, $\eta \geq 15$ and $k \geq 1$. The other positive function has the form:

$$\varphi : [0, \infty) \to [0, \infty), \varphi(s) = \zeta \sqrt[r+1]{\gamma s^r + \beta}, \tag{4.3}$$

with $\zeta, \gamma, \beta \in [1, 4)$ and $r \in (0, 1)$.

The next explicit iterative finite difference-based numerical approximation algorithm was constructed for this model [21]:

$$u_{i,j}^{n+1} = u_{i,j}^{n}(1-\lambda) + u_{i,j}^{0}\lambda$$
$$+\alpha\frac{\varphi\left((u_\sigma)_{i+1,j}^n + (u_\sigma)_{i-1,j}^n + (u_\sigma)_{i,j+1}^n + (u_\sigma)_{i,j-1}^n - 4(u_\sigma)_{i,j}^n\right)}{4}$$
$$\left[\begin{array}{l}\psi\left(\left\|(u_\sigma)_{i+h,j}^n - (u_\sigma)_{i,j}^n\right\|\right)\left(u_{i+h,j}^n - u_{i,j}^n\right) + \psi\left(\left\|(u_\sigma)_{i-h,j}^n - (u_\sigma)_{i,j}^n\right\|\right)\left(u_{i-h,j}^n - u_{i,j}^n\right)\\ +\psi\left(\left\|(u_\sigma)_{i,j+h}^n - (u_\sigma)_{i,j}^n\right\|\right)\left(u_{i,j+h}^n - u_{i,j}^n\right) + \psi\left(\left\|(u_\sigma)_{i,j-h}^n - (u_\sigma)_{i,j}^n\right\|\right)\left(u_{i,j-h}^n - u_{i,j}^n\right)\end{array}\right] \tag{4.4}$$

The scale space is created applying this numerical scheme on current texture u^0 at various iterations. A proper scale-space representation is generated by filtering the texture using the anisotropic diffusion-based model until some properly chosen time moments t and differencing the successive filtered textures. One considers the filtering results at iteration moments $4n, n \in \{0, \ldots, K\}, K \geq 4$. The obtained $\{u^0 - u^4, u^4 - u^8, \ldots, u^{4(K-1)} - u^{4K}\}$ constitutes the scale-space representation with K scales. The image $U_m = u^{4(m-1)} - u^{4m}$ at each scale $m \in \{1, \ldots, K\}$ represents the textural component of the PDE-based decomposition of $u^{4(m-1)}$ and contains the contours of the evolving image u [21].

An effective rotation-invariant texture feature descriptor is determined using this scale space. So, a feature extraction is performed on the current texture at multiple scales, by combining Gray Level Co-occurrence Matrices (GLCM) to 2D circular filters. The co-occurrence matrix of image u computes the occurrences of the pairs of pixels with a specific value and offset in that image [4] and is based on formula:

$$CM_{\Delta x,\Delta y}[u](i,j) = \sum_x\sum_y \begin{cases} 1, \; for\; u(x,y) = i \;\&\; u(x+\Delta x, y+\Delta y) = j \\ 0, \; otherwise \end{cases} \tag{4.5}$$

where $(\Delta x, \Delta y)$ is the offset. If $\{(\Delta x_1, \Delta y_1), \ldots, (\Delta x_p, \Delta y_p)\}$ is a properly selected set of offsets, the sequence of GLCMs corresponding to that image is computed as:

$$CM(u) = \{CM_{\Delta x_1,\Delta y_1}(u), \ldots, CM_{\Delta x_p,\Delta y_p}(u)\} \tag{4.6}$$

A GLCM-based featuring process is then performed on u_0 applying (4.6) to its scale space. One obtains the sequence $\{CM(U_1), CM(U_2), \ldots, CM(U_K)\}$ for it. Then, a circular filtering process is applied to the images of each $CM(U_i), i \in \{1, K\}$.

The 2D circular filters are based on the use of the Gabor filter that is described by the Fourier transform function [21, 22]:

$$H_\sigma(\omega) = e^{-\frac{(\omega - \frac{\pi}{\sigma})(\mu\sigma)^2}{2}} \tag{4.7}$$

where $\frac{\pi}{\sigma}$ and $\mu\sigma$ represents the central frequency and the filter bandwidth. Changing the variable $\omega \to \sqrt{\omega_x^2 + \omega_y^2}$, the next transform is obtained:

$$H_\sigma\left(\omega_x, \omega_y\right) = e^{-\frac{\left(\sqrt{\omega_x^2+\omega_y^2}-\frac{\pi}{\sigma}\right)(\mu\sigma)^2}{2}} \tag{4.8}$$

A bank of M 2D filters with circular frequency response is applied to the GLCM-based images related to the scale space. Each filter is characterized by $\sigma = a^q, q \in \{1, \ldots, M\}$, where $a > 1$. Each $CM_{\Delta x_j, \Delta y_j}(U_i), i = 1, \ldots, K, j = 1, \ldots, p$ is mean normalized first, then convolved to each of the M circular filters of the bank. The $L2$ norm is next computed for each of the M filtered images. So, an M dimensional vector composed of these norms is obtained for each $CM_{\Delta x_j, \Delta y_j}(U_i)$, which means p such M dimensional vectors are computed for U_i. They are concatenated into a pM dimensional 1D feature vector, $V(U_i)$, but 2D $\left[p \times M\right]$ versions of this vector corresponding to the ith scale can be created as well. The K texture feature vectors $V(U_i)$ determined at multiple scales are combined into a final texture feature vector corresponding to u_0. A 1D pMK dimensional feature vector is obtained for it concatenating feature vectors at all scales: $V(u_0) = [V(U_1) \;\; V(U_2) \;\; \ldots \;\; V(U_K)]$. 2D ($[pM \times K]$) or even 3D ($[p \times M \times K]$) forms can be obtained for $V(u_0)$, which represents a robust rotation-invariant and noise-insensitive texture descriptor [21].

A supervised texture classification was applied to these feature vectors. A K-Nearest Neighbour (K-NN) classifier with a voluminous training set containing textures grouped in some known classes was used for this task [18]. Euclidean metric is used to determine the distances between input texture feature vectors and the training feature vectors. While the training textures used in this process were noise-free and not rotated, the input textures could be noisy and rotated in order to evaluate the effectiveness of the technique [21].

This multiscale texture recognition method was tested succesfully on thousands of normal and rotated textures. Voluminous texture databases, like Brodatz album [23] and Kylberg texture dataset [24], were used. The texture classes are displayed in Fig. 4.1. The proposed method achieves high recognition rates. The confusion matrices corresponding to recognition experiments performed on the 2 collections are described in [21]. The obtained classification accuracies are 98.7 and 98.4%, since there are few misclassifications. Its overall classification rate is over 98.5%.

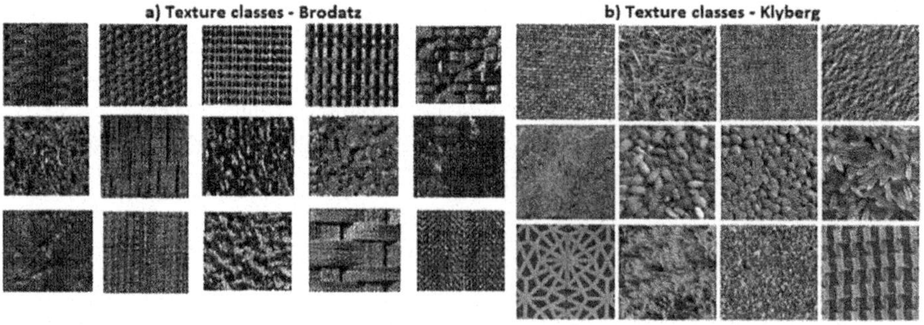

Fig. 4.1 Texture classes related to the 2 datasets

Method comparison that were performed are described in Table 4.1. Our multi-scale texture recognition approach outperforms other well-known texture analysis schemes, including those based on image moments, LBP features, circularly symmetric 2D Gabor filters and CNN-based circular filters, achieving a higher classification accuracy [21].

An unsupervised texture recognition framework was introduced in [25]. The next unsupervised recognition task was considered: the textures from the set $T = \{T_1, \ldots, T_M\}$ has to be clustered automatically in similarity classes whose number is not a priori known, on the basis of their high-level features. We considered the next nonlinear parabolic fourth-order PDE-based model combining second and fourth order anisotropic diffusions for the multi-scale analysis [25].

$$\begin{cases} \frac{\partial u}{\partial t} + \alpha\varphi(\|\nabla u\|)\nabla^2\left(\psi(\|\Delta u_\sigma\|)\nabla^2 u\right) - \beta\varphi(\|\Delta u\|)div(\psi(\|\nabla u_\sigma\|)\nabla u) + \lambda(u - u_0) = 0 \\ u(0, x, y) = u_0(x, y), \quad \forall(x,y) \in \Omega \\ u(t, x, y) = 0, \quad \forall(x,y) \in \partial\Omega; \end{cases} \tag{4.9}$$

where $\alpha, \beta \in [0.5, 1.5)$, $\lambda \in (0, 0.2)$, the positive and monotonic decreasing diffusivity function is $\psi(s) = \gamma\left(\frac{\eta}{|\xi s^2 + \delta|}\right)^{\frac{1}{3}}$, with $\xi \in (0, 1]$, $\delta \geq 5$, $\gamma \geq 1$, $\eta \geq 10$, and $\varphi(s) = \nu(\zeta s^\tau + \varepsilon)^{\frac{1}{\tau+1}}$, where $\varepsilon, \zeta \in [1, 4)$, $\nu \in (1, 1.5)$ and $\tau \in (0, 1)$ [25]. The following finite difference-based numerical approximation scheme that is stable and consistent to the nonlinear PDE model (4.9) is proposed in [25]:

$$\begin{aligned} u_{i,j}^{n+1} &= u_{i,j}^n(1-\lambda) + u_{i,j}^0\lambda - \alpha\varphi\left(\left\|\nabla u_{i,j}\right\|\right)\left(\psi^{i+1,j} + \psi^{i-1,j} + \psi^{i,j+1} + \psi^{i,j-1} - 4\psi^{i,j}\right) \\ &+ \beta\left(\left\|\Delta u_{i,j}\right\|\right)\begin{pmatrix} \psi_{i+\frac{1}{2},j}\left((K_\sigma * u)_{i+1,j}^n - (K_\sigma * u)_{i,j}^n\right) - \psi_{i-\frac{h}{2},j}\left((K_\sigma * u)_{i,j}^n - (K_\sigma * u)_{i-1,j}^n\right) \\ +\psi_{i,j+\frac{1}{2}}\left((K_\sigma * u)_{i,j+1}^n - (K_\sigma * u)_{i,j}^n\right) - \psi_{i,j-\frac{1}{2}}\left((K_\sigma * u)_{i,j}^n - (K_\sigma * u)_{i,j-1}^n\right) \end{pmatrix} \end{aligned} \tag{4.10}$$

where $\psi_{i\pm\frac{h}{2},j} = \frac{\psi_{i\pm h,j} + \psi_{i,j}}{2}$, $\psi_{i,j\pm\frac{h}{2}} = \frac{\psi_{i,j\pm h} + \psi_{i,j}}{2}$ and $K_\sigma(x, y) = \frac{1}{2\pi\sigma^2}e^{-\frac{x^2+y^2}{2\sigma^2}}$.

This numerical approximation algorithm is applied on the texture $u^0 = T_i$. One considers the filtering output at the iteration moments $4k$, where $k \in \{0, \ldots, S\}$, $S \geq 3$. Next, one computes absolute differences between consecutive filtering results and a scale-space representation is obtained as

Table 4.1 Classification performances of several texture analysis methods

Texture analysis technique	Proposed framework (%)	Moment-based approach (%)	Local binary patterns (%)	Circularly symmetric 2D Gabor filters (%)	CNN-based circular filters (%)
Recognition rate	98.5	87	91	97	95

$\{\|u^0 - u^4\|, \|u^4 - u^8\|, \ldots, \|u^{4(S-1)} - u^{4S}\|\}$. The image at each scale $r \in \{1, \ldots, S\}$ is denoted $U_r = u^{4(r-1)} - u^{4r}$. A feature extraction process is then performed on U_r by combining some 2D circular filters [22]. If a bank of Q filters with circular frequency responses is applied to U_r, then a Q-dimensional feature vector, $V(U_r)$, is generated at each scale. The final texture feature vector of the observed texture is determined by concatenating the vectors computed at multiple scales: $V(T_i) = V(u^0) := [V(U_1)\ \ V(U_2)\ \ \ldots\ \ V(U_S)]$ [25].

We proposed a graph-based automatic unsupervised feature vector classification algorithm and applied it on the invariant texture feature vector set $\{V(T_1), \ldots, V(T_M)\}$. The graph-based clustering algorithms have been applied succesfully to texture segmentation. In [25] we proposed a connectivity-based clustering technique adapted for undirected weighted graphs, that is inspired by the Highly Connected Subgraphs (HCS) clustering *algorithm* [26].

Our unsupervised recognition framework was tested successfully on thousands images from voluminous texture collections, such as Kylberg [24] and Brodatz [23] databases. The experiments were performed on textures rotated at various orientations, from those datasets. One used 16 texture classes from each database, which are displayed in Fig. 4.2. Each category of Brodatz collection con-tains 25 [128 × 128] textures for each of the orientations: 0°, 20°, 30°, 45°, 60°, 70°, 90°, 120°. Each category of Kylberg datase contains 20 [576 × 576] textures for each angle.

The described method classifies properly textures rotated by any angle, due to its rotation-invariant multi-scale texture feature extraction approach. It achieves a high texture recognition rate. Our technique obtains high values of Precision, Recall, F1 score and Rand index [27] adapted for pairs of data points, since it generates very few misclassifications. Method comparisons were also performed. It outperforms many unsupervised recognition techniques, including those based on image moments, LBP,

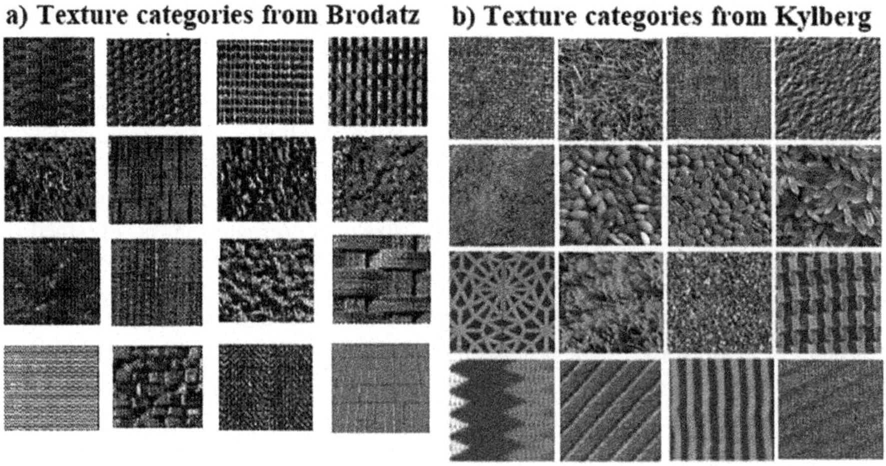

Fig. 4.2 Representative images of the texture classes used in the recognition tests

Table 4.2 Method comparison results

	Precision	Recall	F1
The proposed technique	0.9708	0.9752	0.9730
Proposed multiscale feature extraction + K-means	0.9652	0.9617	0.9634
Proposed multiscale feature extraction + hierarchical agglomerative clustering	0.9638	0.9624	0.9631
Moment-based feature extraction + K-means	0.9268	0.9163	0.9215
Moment-based feature extraction + hierarchical agglomerative clustering	0.9174	0.9197	0.9185
LBP features + K-means	0.9345	0.9367	0.9356
LBP + hierarchical agglomerative clustering	0.9459	0.9422	0.9440
GLCM + K-means	0.9505	0.9574	0.9539
GLCM + hierarchical agglomerative clustering	0.9578	0.9603	0.9590
Gabor 2D filter + K-means	0.9612	0.9608	0.9610
Gabor 2D filter + hierarchical agglomerative clustering	0.9592	0.9611	0.9601

GLCM and 2D Gabor filters in combination to K-means, SOM or hierarchical clustering schemes, in both clean and noisy conditions. Both the nonlinear anisotropic diffusion-based multi-scale texture feature extraction scheme and weighted graph-based texture clustering algorithm proposed here provide better results than other texture analysis and classification methods, as shown by performance metrics values displayed in Table 4.2 [25].

4.2 Content-Based Image Recognition Using Nonlinear PDE-Based Multi-scale Analysis Solutions

Besides the textures, there are also other types of images that can be recognized successfully using the PDE-based multi-scale analysis. Thus, an anisotropic diffusion-based multiscale medical image analysis technique was disseminated by us in [28]. The proposed image analysis framework performs an effective coronavirus disease diagnosis using the Chest X-Ray images [29]. First, our machine learning–based COVID-19 detection technique performs a multi-scale Chest X-ray image analysis using a nonlinear anisotropic diffusion-based scale-space and a 2D circular filter bank. We proposed the following fourth-order anisotropic diffusion model with boundary conditions to create that scale space representation:

$$\begin{cases} \frac{\partial u}{\partial t} + \lambda\psi(|\nabla u_\sigma|)\Delta\left(\varphi\left(\left\|\nabla^2 u\right\|\right)\Delta u\right) + \delta(u - u_0) = 0 \\ u(0, x, y) = u_0(x, y), \quad \forall (x,y) \in \Omega \\ u(t, x, y) = 0, \quad \forall (x,y) \in \partial\Omega; \end{cases} \tag{4.11}$$

where the conductance function $\varphi(s) = \alpha\sqrt[3]{\frac{\beta}{|\gamma s^2+\eta|}}$, with $\gamma \in (0,1], \alpha \geq 1, \eta, \beta \geq 5$, and $\psi(s) = \zeta(\nu s^r + \rho)^{\frac{1}{r+1}}$, with $\zeta, \nu, \rho \in [1, 5)$ [28].

The nonlinear PDE model is non-variational, since it is not derived from an energy functional minimization, and also well-posed, since it admits a unique variational (weak) solution that is determined applying a numerical solving scheme. Since the PDE can be written as $\frac{\partial u}{\partial t} + \delta(u - u_0) = -\lambda\psi(|\nabla u_\sigma|)\Delta\left(\varphi\left(\left\|\nabla^2 u\right\|\right)\Delta u\right)$, its left term can be discretized using central differences as:

$$\frac{u_{i,j}^{n+\Delta t} - u_{i,j}^n}{\Delta t} + \delta\left(u_{i,j}^n - u_{i,j}^0\right) = u_{i,j}^{n+\Delta t}\frac{1}{\Delta t} + u_{i,j}^n\left(\delta - \frac{1}{\Delta t}\right) - u_{i,j}^0\delta \tag{4.12}$$

where $x = ih, y = jh, i \in \{1, \ldots, I\}, j \in \{1, \ldots, J\}, t = n\Delta t, n \in \{0, \ldots, N\}$.

The right term $\lambda\psi(|\nabla u_\sigma|)$ is approximated as $\lambda\psi\left(\nabla\left\|(K_\sigma * u)_{i,j}\right\|\right)$ and $\Delta\left(\varphi\left(\left\|\nabla^2 u\right\|\right)\Delta u\right)$ is discretized as:

$$\Delta\varphi_{i,j} = \frac{\varphi_{i+h,j} + \varphi_{i-h,j} + \varphi_{i,j+h} + \varphi_{i,j-h} - 4\varphi_{i,j}}{h^2} \tag{4.13}$$

where $\Delta u_{i,j} = \nabla^2 u_{i,j} = \frac{u_{i+h,j}^n + u_{i-h,j}^n + u_{i,j+h}^n + u_{i,j-h}^n - 4u_{i,j}}{h^2}$ and $\varphi_{i,j} = \varphi\left(\left|\Delta u_{i,j}\right|\right)\nabla^2 u_{i,j}$. We take $h = \Delta t = 1$ and obtain the following numerical approximation scheme:

$$\begin{aligned} u_{i,j}^{n+1} &= u_{i,j}^n(1 - \delta) + u_{i,j}^0\delta \\ &\quad - \lambda\psi\left(\left\|\nabla(K_\sigma * u)_{i,j}\right\|\right)\left(\varphi_{i+1,j} + \varphi_{i-1,j} + \varphi_{i,j+1} + \varphi_{i,j-1} - 4\varphi_{i,j}\right) \end{aligned} \tag{4.14}$$

This iterative numerical discretization algorithm is consistent to the PDE model and converges to its fast to its numerical solution. The following scale-space representation with K scales is obtained by applying this numerical scheme: $\left\{\left\|u^0 - u^5\right\|, \left\|u^5 - u^{10}\right\|, \ldots, \left\|u^{5(K-1)} - u^{5K}\right\|\right\}$. A rotation-invariant feature extraction process is performed at each scale. A 2D band-pass circularly symmetric filter is considered for this task. The image at each kth scale, $\left\|u^{5(k-1)} - u^{5k}\right\|$, is mean-normalized and convolved to a set of M 2D circular filters having the form $F_\sigma\left(\omega_x, \omega_y\right) = e^{-\frac{\left(\sqrt{\omega_x^2+\omega_y^2} - \frac{\pi}{\sigma}\right)(\mu\sigma)^2}{2}}$ [28]. A M-dimensional feature vector that is $V\left(\left\|u^{5(k-1)} - u^{5k}\right\|\right)$ is achieved at each scale. Final feature vector of the observation, representing a powerful rotation-invariant MK-dimensional descriptor of the Chest X-ray image is determined concatenating the vectors obtained at multiple scales.

A supervised binary classification is applied to these medical image feature vectors. One considers a voluminous medical training set composed of Chest X-ray images divided into 2 classes: first one contains normal images, related to healthy subjects, while the second one contains the Chest X-ray images of infected subjects. See examples of normal and COVID-19 infection related images in Fig. 4.3.

A K-Nearest Neighbour (K-NN) classifier based on this training set is then applied to the medical images to be recognized. Each input Chest X-ray image is thus recognized as either normal or coronavirus positive image [28].

This recognition technique was tested on a several hundreds of Chest X-Ray images from the Kaggle repository [30]. A training set of 115 normal Chest X-Ray images and 115 COVID-19 positive Chest X-Ray images was created. We also considered a testing set composed of 65 normal Chest X-ray images and 65 COVID-19 positive images [28]. The multi-scale feature extraction and K-NN classifier were applied to the Chest X-ray images with parameters: $K = 4$ scales, $M = 4$ circular filters and $K_{NN} = 7$. A high recognition accuracy rate (over 92%) was achieved. Its effectiveness is illustrated by the method comparison results described in Table 4.3.

Table 4.3 Method comparison

Technique	Precision	Recall	F1	Accuracy
Proposed model	92.97	89.76	91.34	92.12
DarkCovidNet	89.96	85.35	87.37	87.02
Flat-EfficientNet B3	93.93	93.96	93.94	93.34
Covid-Net	98.9	91	94.79	93.3

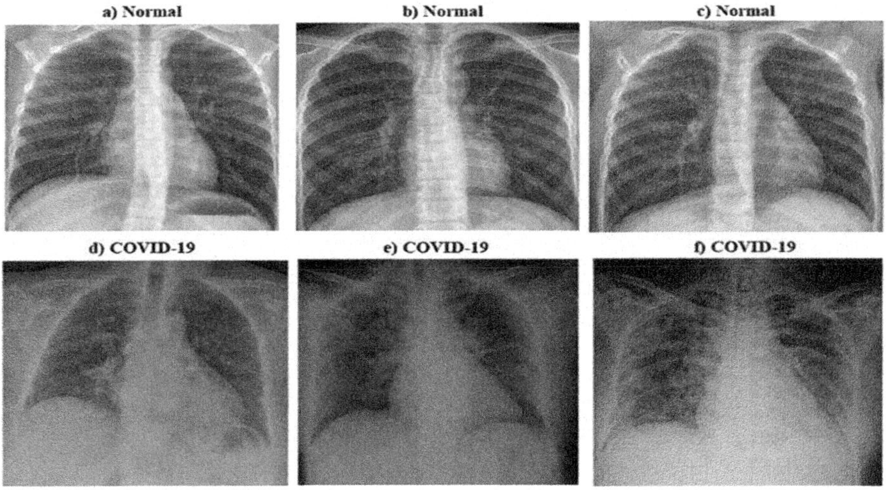

Fig. 4.3 Examples of normal and COVID-19 infection related images

4.3 Image Indexing and Retrieval Frameworks Using Nonlinear Diffusion-Based Multi-scale Image Analysis

The nonlinear PDE-based multi-scale analysis has also be used sucessfully in the content-based image indexing and retrieval domain. The content-based image retrieval (CBIR) techniques retrieve the images on the basis of their content features instead of textual descriptions [31]. Their querying process uses the *query by example*, which means providing user-supplied query images or image content features. The features involved in the retrieval process can be *global* features, based on color, texture, shape and spatial information, or *local* features. The global feature-based CBIR systems use invariant color moments [32], histograms, correlograms [31] and co-occurrence matrices [33], Wavelet transforms [34], gray-level co-occurrence matrices (GLCM) [35], Gabor filters [36], Fourier descriptors [37], affine moment invariants [38], local structure descriptors [31] and color, texture and shape feature fusion [39]. Local feature-based image retrieval is based on scale-invariant feature transform (SIFT) [40], binary robust invariant scalable key points (BRISK) [31], speeded-up robust features (SURF) [41], local binary patterns (LBP) [42] and Harris corner detectors [31].

Various image content-based database indexing structures that facilitate the retrieval tasks have been developed, many of them based on clustering techniques. These approaches create cluster-based indexing structures by clustering the image features based on similarity, using various machine learning algorithms [18]. A CBIR framework using a cluster-based image indexing model created using the nonlinear PDE-based multi-scale analysis was introduced by us in [43].

The scale-space representation was obtained by considering the next nonlinear second-order anisotropic diffusion-based model:

$$u_{i,j}^{n+1} = u_{i,j}^{n}(1-\beta) + u_{i,j}^{0}\beta + \lambda\eta_{i,j}\Big(\psi_{i+\frac{1}{2},j}\Big(u_{i+1,j}^{n} - u_{i,j}^{n}\Big) - \psi_{i-\frac{h}{2},j}\Big(u_{i,j}^{n} - u_{i-1,j}^{n}\Big) + \psi_{i,j+\frac{1}{2}}\Big(u_{i,j+1}^{n} - u_{i,j}^{n}\Big) - \psi_{i,j-\frac{1}{2}}\Big(u_{i,j}^{n} - u_{i,j-1}^{n}\Big)\Big) \quad (4.15)$$

where $\alpha \in [1, 2)$, $\beta \in (0, 0.4)$, $\psi(s) = \gamma\left(\frac{\alpha}{|\xi s^2+\delta|}\right)^{1/3}$, $\xi \in (0, 1]$, $\gamma \geq 1$, $\alpha, \delta \geq 4$, $\eta(s) = \sqrt[\kappa+1]{\frac{\nu s^\tau+\rho}{\zeta}}$, $\nu, \rho \in [1, 4)$, $\zeta \in (0, 0.6)$ and $\kappa \in (0, 1)$.

This second-order nonlinear PDE model is non-variational and well-posed, since it exists a unique weak solution for it. That solution is determined applying the finite difference-based numerical approximation scheme proposed in [43]:

$$u_{i,j}^{n+1} = u_{i,j}^{n}(1-\beta) + u_{i,j}^{0}\beta + \lambda\eta_{i,j}\Big(\psi_{i+\frac{1}{2},j}\Big(u_{i+1,j}^{n} - u_{i,j}^{n}\Big) - \psi_{i-\frac{h}{2},j}\Big(u_{i,j}^{n} - u_{i-1,j}^{n}\Big) + \psi_{i,j+\frac{1}{2}}\Big(u_{i,j+1}^{n} - u_{i,j}^{n}\Big) - \psi_{i,j-\frac{1}{2}}\Big(u_{i,j}^{n} - u_{i,j-1}^{n}\Big)\Big) \quad (4.16)$$

where space and time coordinates are quantized as $x = ih$, $y = jh$, $i \in \{1, \ldots, I\}, j \in \{1, \ldots, J\}, t = n\Delta t, n \in \{0, \ldots, N\}$ and $\psi_{i\pm\frac{h}{2},j} = \frac{\psi_{i\pm h,j}+\psi_{i,j}}{2}$, $\psi_{i,j\pm\frac{h}{2}} = \frac{\psi_{i,j\pm h}+\psi_{i,j}}{2}$ [43].

It was then successfully used to create a scale-space representation. The diffusion model works for grayscale images, while one needs a multi-scale space for RGB color images. Applying (4.16) on each of the 3 image channels is not a proper solution, since R, G and B channels could be highly correlated. So, the RGB image *Im* was converted to a decorrelated color space like CIELAB, then its luminance channel *L(Im)* was processed by (4.16), $(L(Im))^n$ representing its filtering after *n* iterations. The filtered L*a*b image, $\left[(L(Im))^n, a(Im), b(Im)\right]$, is then converted back to RGB form. One obtains next multiscale representation with *K* scales for *Im*

$$S(Im) = \left\{Im, RGB\left(\left[(L(Im))^{\tau}, a(Im), b(Im)\right]\right), \ldots, RGB\left(\left[(L(Im))^{\tau(K-1)}, a(Im), b(Im)\right]\right)\right\} \tag{4.17}$$

where $K \geq 5$ and time step $\tau \in [3, 10]$ [43]. The scale space *S* is then used in the multi-scale image content feature extraction process. A high-level feature extraction is performed on the analyzed image at each scale $k \in \{0, \ldots, K-1\}$, having the form $S(Im)\{k\} = RGB\left(\left[(L(Im))^{\tau k}, a(Im), b(Im)\right]\right)$, using deep learning-based models [43].

Convolutional Neural Networks (CNNs, ConvNets) represent deep learning networks that provide powerful image content feature extraction results [44]. A CNN learns successfully high-level feature representations for a large variety of images if it has been trained on voluminous databases. Here one applied 2 pre-trained CNNs to compute the high-level characteristics of each image: ResNet (Residual Network) 50 and Inception-V3. ResNet-50 is a deep residual network characterized by a 50-layer architecture trained on 1.2 million images of the voluminous ImageNet database containing 1000 object categories [45], while Inception-V3 is a 48 layers deep convolutional neural network that started as a GoogleNet module [46]. The layers of these ConvNets generate activations to the input images. While their first layers can detect only low-level characteristics that are then processed by the deeper convolutional layers that combine them to achieve high-level features [45, 46]. We considered as optimal content-based feature extraction solutions the Fully Connected Layer with 1000 neurons (FC 1000), located right before final classification layer in the ResNet-50 architecture, and Predictions Layer of Inception-V3 that is also a fully connected layer located before classification layer. Each $S(Im)\{k\}$ image was pre-processed according to the specifications of ResNet-50 input layer, first. It was resized at the [224 × 224 × 3] required format and a *0*-center normalization was performed to it: $S(Im)\{k\} := (S(Im)\{k\} - \mu(S(Im)\{k\}))/\sigma(S(Im)\{k\}), \forall k \in \{0, \ldots, K-1\}$. The mini-batch parameter size was set at 32. Next, these pre-processed RGB images are fed into the ResNet-50 network. For each *k* from 0 to $K - 1$, its FC 1000 layer produces an activation on $S(Im)\{k\}$ that computes high-level characteristics in the form of a content feature vector with 1000 coefficients, $V_1(S(Im)\{k\})$. The Inception network was then applied similarly on the initial $S(Im)\{k\}$. The image is resized at the [299 × 299 × 3] format required by its input layer, then the same normalization is performed to it. It is fed into Inception-V3 model whose fully connected *prediction* layer produces an activation generating 1000 coefficient feature vector $V_2(S(Im)\{k\})$. All 2D feature vectors $V(S(Im)\{k\}) := [V_1(S(Im)\{k\}); V_2(S(Im)\{k\})]$ of the scale space were concatenated into the final $[2 \times 1000K]$ feature vector $V(Im) := [V(S(Im)\{0\}) \ldots V(S(Im)\{K-1\})]$, that is a powerful content descriptor of Im [43].

A content-based database indexing structure was then created applying an automatic graph-based image feature vector clustering approach [43]. A weighted similarity graph was modeled for image database $Db = \{Im_1, \ldots, Im_M\}$, as a fully connected undirected graph $G = (V, E)$, whose vertex set V corresponds to images Im_i and $E \subseteq V \times V$ represents the set of linking edges weighted by this weighting function:

$$w : E \rightarrow R : w_{ij} = w(i,j) = 1000e^{-d\left(V(Im_i), V(Im_j)\right)}, \quad \forall i,j \in \{1, \ldots, M\}, \quad i \neq j \tag{4.18}$$

where d is the Euclidian metric. A normalized-cut (NCut) based graph clustering algorithm was provided for G in [47]. The achieved graph clustering result $C = \{C\{1\}, \ldots, C\{N_c\}\}$, where N_c is the detected number of clusters, was then used as a cluster-based index of *Db*. The image coresponding to the maximum sum of within-cluster weights was selected as representative for each $C\{r\}$ cluster. So, that image is $Im_{ind(r)} \in Db$, where $ind(r) := \arg\max_{i \in C\{r\}} \sum_{j \in C\{r\}} w_{ij}$ [43].

The query Q received as input by the proposed CBIR system is compared to its representative images only. The most relevant one, which is $Im_{ind(j)}$ with $j = \arg\min_{i \in \{1,\ldots,N_c\}} d\big(V(Q), V\big(Im_{ind(i)}\big)\big)$ is provided to the user if it is not too dissimilar to the query, so the condition $d\big(V(Q), V\big(Im_{ind(j)}\big)\big) \leq 2\frac{\sum_{i,k \in \{1,\ldots,N_c\}} d\left(V\left(Im_{ind(i)}\right), V\left(Im_{ind(k)}\right)\right)}{N_c(N_c-1)}$ holds.

A relevance-feedback mechanism is then used to search for better retrieval results in the database index given by the most relevant cluster $C\{j\}$. The most relevant R images to Q are retrieved from $C\{j\}$ and ranked according to their relevance given by feature vector distance value $d(V(Q), V(Im_i)), i \in C\{j\}$. All or some of them can be extracted as retrieval output and *the best* of them (usually the 1st-ranked one) may be selected as a new query Q for the retrieval process [43].

This retrieval framework combining successfully PDE to AI models was tested on ODIDS (*OutDoor Images Data Set*) database containing over 5000 color images, grouped into 43 categories [48]. We created a validation dataset based on it, containing 1750 images belonging to 34 similarity classes and 80 images not belonging to any of them. We used 1300 of the class-belonging images as database images to be retrieved and the remaining 450 as first queries. The 80 no-class images were also used for queries. A CBIR test is partially described (only 1 query) in Fig. 4.4.

This framework was compared to other cluster-based CBIR systems using global features, like color moment invariants, GLCM and 2D Gabor filters, combined global features, or local features, such as LBP, SURF, SIFT or MKSIFT [49], with clustering

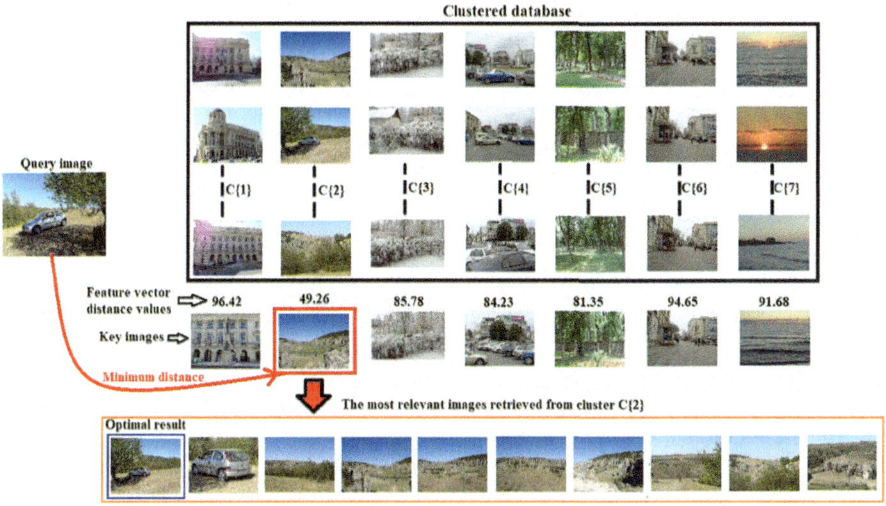

Fig. 4.4 Example of a CBIR experiment

Table 4.4 Method comparison: MAP values achieved by various cluster-based CBIR models

CBIR technique	Mean average precision (MAP)
The proposed framework	0.9384
Color moment invariant + hierarchical agglomerative clustering	0.6471
GLCM + *K*-means	0.6843
SIFT + *K*-means	0.7302
MKSIFT + Bayesian fuzzy clustering	0.9453
Color, shape, texture features + agglomerative clustering scheme	0.8697
LBP features + *K*-means	0.7815
SURF + hierarchical agglomerative clustering	0.7521
2D Gabor filters + hierarchical agglomerative clustering	0.8162

algorithms like *K*-means, hierarchical agglomerative clustering, Bayesian fuzzy clustering or other ML schemes. It outperforms most of them and is slightly outperformed only by MKSIFT + BFC, as illustrated by method comparison results displayed in Table 4.4.

References

1. Z. Boulkenafet, J. Komulained, X. Feng, A. Hadid, Scale space texture analysis for face anti-spoofing, *2016 International Conference on Biometrics (ICB)*, pp. 1–6, June 2016.
2. A. Fernandez, M. X. Alvarez, F. Bianconi, Texture description through histograms of equivalent patterns, *J. Math. Imag. Vis.*, Vol. 45, No. 1, pp. 76–102, Jan. 2013.
3. M. Tuceryan, Moment-based texture segmentation, *Pattern recognition letters*, Vol. 15, No. 7, pp. 659–668, 1994.
4. R. M. Haralick, K. Shanmugan, Textural features for image classification, *IEEE Transactions on systems, man, and cybernetics*, 6, pp. 610–621, 1973.
5. T. Ojala, M. Pietikainen, D. Harwood, A comparative study of texture measures with classification based on featured distributions, *Pattern recognition*, 29, 1, pp. 51–59, 1996.
6. L. Zhang, Z. Zhou, H. Li, Binary Gabor pattern: An efficient and robust descriptor for texture classification, *19th IEEE international conference on image processing*, pp. 81–84, Sept. 2012.
7. F. Ershad, Texture classification approach based on energy variation, *International Journal of Multimedia Technology*, 2 (2), pp. 52–55, 2012.
8. H. Liang, D. S. Weller, Edge-based texture granularity detection, *2016 IEEE International Conference on Image Processing (ICIP)*, pp. 3563–3567, Sept. 2016.
9. I. Epifanio, P. Soille, Morphological texture features for unsupervised and supervised segmentations of natural landscapes, *IEEE Transactions on Geoscience and Remote Sensing*, 45, 4, pp. 1074–1083, 2007.
10. D. G. Lowe, Distinctive image features from scale-invariant keypoints, *International journal of computer vision*, vol. 60, no. 2, pp. 91–110, 2004.
11. Y. Xu, X. Yang, H. Ling, H. Ji, A new texture descriptor using multifractal analysis in multi-orientation wavelet pyramid, *IEEE Computer Society Conference on Computer Vision and Pattern Recognition*, pp. 161–168, June 2010.
12. G. R. Cross, A. K. Jain, Markov random field texture models, *IEEE Transactions on Pattern Analysis & Machine Intelligence*, 1, pp. 25–39, 1983.
13. J. Mao, A. K. Jain, Texture classification and segmentation using multiresolution simultaneous autoregressive models, *Pattern recognition*, 25, 2, pp. 173–188, 1992.
14. A. K. Jain, F. Farokhnia, Unsupervised texture segmentation using Gabor filters, *Pattern recognition*, 24, 12, pp. 1167–1186, 1991.
15. S. Livens, P. Scheunders, G. Wouver, D. van Dyck, *Wavelets for texture analysis, an overview*, pp. 581–585, 1997.
16. L. Shen, Q. Yin, Texture classification using curvelet transform, *Proceedings of the International Symposium on Information Processing (ISIP 2009)*, pp. 319, Academy Publisher, 2009.
17. P. Yang, G. Yang, Feature extraction using dual-tree complex wavelet transform and gray level co-occurrence matrix, *Neurocomputing*, 197, pp. 212–220, 2016.
18. C. M. Bishop, *Pattern recognition and machine learning*, Springer, 2006.
19. A. Vincent, P. F. Whelan, Using filter banks in convolutional neural networks for texture classification, *Pattern Recognition Letters*, 84, pp. 63–69, 2016.
20. S. E. Schaeffer, Graph clustering, *Comput. Sci. Rev.*, 1, pp. 27–64, 2007.
21. T. Barbu, P. Ungureanu, L. Goraş, Multiscale Texture Recognition using Anisotropic Diffusion-based Scale Space and Combined Rotation-invariant Feature Descriptors, *Proc. Rom. Acad. Ser. A*, Vol. 21, pp. 385–393, 2020.
22. R. Porter, N. Canagarajah, Gabor filters for rotation invariant texture classification, *Proceedings of the IEEE Intl. Symp. on Circuits and Systems*, H.K, China, 9–12 June, pp. 1193–1196, 1997
23. P. Brodatz, *Textures: A Photographic Album for Artists and Designers*, Dover: New York, 1966.
24. G. Kylberg, I. Sintorn, On the influence of interpolation method on rotation invariance in texture recognition, *J Image Video Proc.* 2016, 17, 2016.
25. T. Barbu,Automatic unsupervised texture recognition framework using anisotropic diffusion-based multi-scale analysis and weight-connected graph clustering, *Symmetry*, *13*(6), 925, 2021.

26. B. Bollobás, *Modern Graph Theory*, Springer: Berlin/Heidelberg, Germany, 2013; Volume 184.
27. D. M. Powers, Evaluation: From precision, recall and F-measure to ROC, informedness, markedness and correlation, *arXiv preprint* arXiv:2010.16061.
28. T. Barbu, M. Luca, A. Ciobanu, Anisotropic Diffusion-based Multiscale Medical Image Analysis Technique for COVID-19 Detection, *2021 International Conference on e-Health and Bioengineering (EHB)*, pp. 1–4, 2021 IEEE.
29. A. J. Rodriguez-Morales et al., Clinical, laboratory and imaging features of COVID-19: A systematic review and meta-analysis, *Travel medicine and infectious disease*, 34, 2020.
30. Kaggle, *COVID-19 chest xray*, https://www.kaggle.com/bachrr/covid-chest-xray
31. I. M. Hameed, S. H. Abdulhussain, B. M. Mahmmod, Content-based image retrieval: A review of recent trends, *Cogent Engineering*, 8.1, Article 1927469, June 2021.
32. X. Duanmu, Image retrieval using color moment invariant, *2010 Seventh International Conference on Information Technology: New Generations*, pp. 200–203, Las Vegas, Nevada, 2010.
33. G. Qiu, Color image indexing using BTC, *IEEE Transactions on Image Processing*, 12 (1), pp. 93–101, January 2003.
34. R. Ashraf, M. Ahmedm, S. Jabbar, S. Khalid, A. Ahmad, S. Din, G. Jeon, Content based image retrieval by using color descriptor and discrete Wavelet transform, *Journal of Medical Systems*, 42 (3), 44, March 2018.
35. B. Ramamurthy, K. R. Chandran, Content Based Medical Image Retrieval with Texture Content Using Gray Level Co-occurrence Matrix and K-Means Clustering Algorithms, *Journal of Computer Science*, 8 (7), pp. 1070–1076, 2012.
36. D. Zhang, A. Wong, M. Indrawan, G. Lu, Content-based image retrieval using Gabor texture features, *IEEE Transactions Pami*, 3656, 2000.
37. D. Zhang, G. Lu, A comparative study on shape retrieval using Fourier descriptors with different shape signatures, *Proceedings of the International Conference on Intelligent Multimedia and Distance Education*, pp. 1–9, 2001.
38. T. Suk, J. Flusser, Affine moment invariants generated by graph method, *Pattern Recognition*, 40 (2), pp. 2047–2056, 2011.
39. S. Pandey, P. Khanna, Content-based image retrieval embedded with agglomerative clustering built on information loss, *Computers & Electrical Engineering*, 54, pp. 506–521, 2016.
40. G. A. Montazer, D. Giveki, Content based image retrieval system using clustered scale invariant feature transforms, *Optik* (Stuttg), 126 (18), pp. 1695–1699, Sept. 2015.
41. H. Bay, A. Ess, T. Tutelaars, L. Van Gool, Speeded-Up Robust Features (SURF), *Computer Vision and Image Understanding*, 110 (3), pp. 346–359, 2008.
42. D. Kulshreshtha, V. Singh, A. Shrivastava, A. Chaudhary, R. Srivastava, Content-based Mammogram Retrieval Using k-Means Clustering and Local Binary Pattern, *Proc. ICIVC 2017*, Chengdu, China, pp. 634–638, 2–4 June 2017.
43. T. Barbu, Content-based Image Retrieval Framework using Multi-scale Deep Learning – based Feature Extraction and Graph-based Clustering, *Proceedings of the Romanian Academy,* Series A, Volume 23, Number 3, pp. 287–296, July-September 2022.
44. J. Wan, D. Wang, S. C. H. Hoi, P. Wu, J. Zhu, Y. Zhang, J. Li, Deep learning for content-based image retrieval: A comprehensive study, *Proceedings of the 22nd ACM international conference on Multimedia*, pp. 157–166, 2014.
45. K. He, X. Zhang, S. Ren, J. Sun, Deep residual learning for image recognition, *Proceedings of the IEEE conference on computer vision and pattern recognition*, CVPR 2016, pp. 770–778, 2016.
46. C. Szegedy, V. Vanoucke, S. Ioffe, J. Shlens, Z. Wojna, Rethinking the inception architecture for computer vision, *Proceedings of the IEEE Conference on Computer Vision and Pattern Recognition*, pp. 2818–2826. 2016.
47. J. Shi, J. Malik, Normalized cuts and image segmentation, *IEEE Transactions on pattern analysis and machine intelligence*, 22.8, pp. 888–905, 2000.

48. A. Ignat, M. Luca, I. Păvăloi, C. L. Lazăr, DENOL: efficient descriptors selection for automatic image retrieval, *SOFA 2020 - 9th International Workshop on Soft Computing Applications*, Arad, Romania, 27–29 Nov. 2020, http://odids.iit.academiaromana-is.ro
49. B. M. Kumar, R. Pushpalakshmi, An approach for image search and retrieval by cluster-based indexing of binary MKSIFT codes, *The Computer Journal*, 63.6, pp. 857–879, 2020.

Chapter 5
Variational and PDE-Based Static and Video Image Segmentation Approaches

This chapter approaches the PDE-based image segmentation domain, describing variational and non-variational PDE-based models for static and video image segmentation. The edge-based segmentation of static images is addressed in the first section of the chapter, where our contribution in this field, representing a nonlinear diffusion-based multi-scale edge detection framework, is discussed. Next, the segmentation solutions using nonlinear PDE-based parametric and geometric active contour models are presented in the second section. Our own contributions, representing some geodesic active contours with level-set functions, are also described here. The video image segmentation field is then addressed in the third section. A temporal video segmentation technique proposed by us, which combines a nonlinear second-order PDE-based multi-scale frame analysis to a deep learning-based high-level feature extraction, is discussed in that section.

5.1 Diffusion-Based Multi-scale Analysis for Edge-Based Segmentation

The edge detection represents an important image analysis field that includes techniques identifying the boundaries, which contain the points of a image where its brightness has sharp discontinuities. The edges correspond to the discontinuities in depth or surface orientation, variations in scene illumination or changes in material properties [1]. The application areas of this domain include object detection and recognition, biometric authentication, image compression and remote sensing.

Various edge detectors have been developed in the last decades. They may be grouped into two major categories. The first one includes techniques based on the first-order image derivative [2], like classical gradient operators of Roberts, Sobel and Prewitt. The second category contains detection methods using second-order derivative [3], such as the Laplacian of a Gaussian (LoG)-based edge operator of

T. Barbu, *Digital Image Processing, Analysis and Computer Vision Using Nonlinear Partial Differential Equations*, Studies in Computational Intelligence 1211,
https://doi.org/10.1007/978-3-031-89576-0_5

Marr and Hildreth [4]. The most known detector, still representing the state-of-the-art of the area, is the multi-stage Canny edge detection algorithm of Canny [5].

Since the edges represent multi-scales structures in nature and human vision has a multi-scale character too, numerous multi-scale edge detection solutions have been introduced [6]. The multi-scale representations provide better boundary extraction solutions in natural scenes and work properly in noisy conditions. The early multi-scale edge detection techniques applied the 2D Gaussian filtering at multiple scales provided by σ, and fine-to-coarse and coarse-to-fine boundary tracking operations [7, 8]. Edge focusing methods represent more performant techniques that extract the edges in a image filtered by a large scale Gaussian filter, using the adaptive thresholding, and next combine the edge information by moving from a coarse to a fine scale [9]. Another improved multiscale boundary detection approach, introduced by Williams and Shah [10], use the motion information of the edges filtered by a Gaussian operator of different sizes, in order to determine how to link the edge points detected at multiple scales. An extended multi-scale version of the single-scale Sobel detector, based on 2D Gaussian 2D filtering and fine-to-coarse tracking, is disseminated in [11]. The multi-scale anisotropic Gaussian kernels (AGKs) represent another effective edge detection tool [12]. Various multi-scale edge identification methods using other image filters have been also developed. Some effective multi-scale detection approaches are based on Wavelet transforms [13, 14].

Nonlinear partial differential equation-based filters have also been used successfully to create effective multi-scale edge detection solutions. The influential anisotropic diffusion-based model developed by Perona and Malik was applied to create a scale-space representation for image boundary extraction [15].

Some nonlinear PDE-based edge-based segmentation frameworks have been also developed by us [16, 17]. Thus, a second-order nonlinear anisotropic diffusion-based edge detection model proposed by us is disseminated in [16]. It was already described in the Sect. 3.2 of Chap. 3, since it is part of the nonlinear PDE-based image compression framework described there.

Another nonlinear diffusion-based edge detection technique introduced by us is disseminated in [17]. The next second-order parabolic PDE-based filtering model is introduced there to create the scale-space representation:

$$\begin{cases} \frac{\partial u}{\partial t} - \beta\delta(\|\nabla u\|)div(\varphi(\|\nabla u_\sigma\|)\nabla u) + \alpha(u - u_0) = 0, & \forall(x,y) \in \Omega \\ u(0,x,y) = u_0(x,y), & \forall(x,y) \in \Omega \\ u(t,x,y) = 0, & \forall(x,y) \in \partial\Omega; \end{cases} \tag{5.1}$$

where $\alpha \in (0, 0.3)$, $\beta \in [1, 2)$, $u_0 \in L^2(\Omega)$, $u_\sigma = u * G_\sigma$, $G_\sigma(x, y) = \frac{1}{2\pi\sigma^2}e^{-\frac{x^2+y^2}{2\sigma^2}}$, the proposed positive, monotonic decreasing diffusivity function has the form

$$\varphi : [0, \infty) \to [0, \infty) : \varphi(s) = \xi \sqrt[3]{\frac{\lambda}{|\eta s^2 + \gamma|}} \tag{5.2}$$

where $\eta \in (0, 1]$, $\gamma \geq 4$, $\xi \geq 1$, $\lambda \geq 5$ and the other positive function of the model is

$$\delta : [0, \infty) \to [0, \infty) : \delta(s) = \left(\frac{\varepsilon s^{\tau} + \zeta}{\nu}\right)^{\frac{1}{\tau+1}}, \tag{5.3}$$

where $\varepsilon, \zeta \in [1, 5)$, $\nu \in (0, 0.5)$ and $\tau \in (0, 1)$.

This second-order diffusion model is non-variational and well-posed, since it admits a unique and weak solution. This solution is determined numerically by applying an iterative approximation algorithm for (5.1) that converges to it. The proposed discretization scheme which solves numerically our PDE model is detailed in [17]. It has been obtained by applying the finite difference method and has the following explicit form:

$$\begin{aligned} u_{i,j}^{n+1} &= u_{i,j}^{n}(1-\alpha) + u_{i,j}^{0}\alpha \\ &+ \beta\delta_{i,j}\begin{pmatrix} \varphi_{i+\frac{1}{2},j}\left((G_\sigma * u)_{i+1,j}^{n} - (G_\sigma * u)_{i,j}^{n}\right) - \varphi_{i-\frac{h}{2},j}\left((G_\sigma * u)_{i,j}^{n} - (G_\sigma * u)_{i-1,j}^{n}\right) \\ +\varphi_{i,j+\frac{1}{2}}\left((G_\sigma * u)_{i,j+1}^{n} - (G_\sigma * u)_{i,j}^{n}\right) - \varphi_{i,j-\frac{1}{2}}\left((G_\sigma * u)_{i,j}^{n} - (G_\sigma * u)_{i,j-1}^{n}\right) \end{pmatrix} \end{aligned} \tag{5.4}$$

where $x = ih$, $y = jh$, $i \in \{1, \ldots, I\}$, $j \in \{1, \ldots, J\}$, $t = n\Delta t$, $n \in \{0, \ldots, N\}$ and

$$\varphi_{i\pm\frac{h}{2},j} = \frac{\varphi_{i\pm h,j} + \varphi_{i,j}}{2}, \varphi_{i,j\pm\frac{h}{2}} = \frac{\varphi_{i,j\pm h} + \varphi_{i,j}}{2} \tag{5.5}$$

This numerical approximation algorithm is stable and consistent to (5.1) and converges in N iterations to its variational solution. It has been successfully used to create an effective multiscale space for the edge detection task [17]. A multi-scale representation with K scales is then generated by considering the output of this numerical scheme at K iteration moments. The scale-space is determined as following:

$$S = \left\{\left|u^{\tau} - u^{0}\right|, \left|u^{2\tau} - u^{0}\right|, \ldots, \left|u^{K\tau} - u^{0}\right|\right\} \tag{5.6}$$

where $K \geq 10$, the time step $\tau \in [30, 50]$ and $k \in \{1, \ldots, K\}$ [17].

The image at each kth scale of S, denoted as $U^k = \left|u^{k\tau} - u^0\right|$, is then analyzed to detect its edges. One determines the zero-crossings of U^k and the binary image representing them is denoted $E\left(U^k\right)$. It is further processed, only the zero-crossings corresponding to those pixels of U^k characterized by a gradient magnitude exceeding a given threshold being kept in $E\left(U^k\right)$. This process is formulated as follows:

$$\left\|\nabla U_{ij}^{k}\right\| \leq T_k \Rightarrow E\left(U^k\right)[i, j] := 0, \ \forall k \in \{1, \ldots, K\}, \ i \in \{1, \ldots, I\}, \ j \in \{1, \ldots, J\} \tag{5.7}$$

where a proper threshold value is $T_k = 2\mu\left(\left\|\nabla U^k\right\|\right)$ [17]. Since only the real edges must be extracted, this technique discards the connected components of $E\left(U^k\right)$ that may represent noise, their areas being smaller than threshold value of 25 [17]. Some morphological operations are next applied to the obtained binary image. An image closing operation, is applied first: $E(U^k) := (E(U^k) \oplus Sq)\Theta Sq$, where Sq represents a [1 × 1] square structuring element. A thinning process is performed on the closed image applying the Zhang-Suen algorithm [18]. The thinned $E(U^k)$ represents the boundary detection result at the kth scale of S.

The edges extracted at multiple scales are combined into the final detection result by applying a fine-to-coarse edge tracking technique. At lower scales, provided by low k values, the edge maps $E(U^k)$ correspond to finer details and some of white pixels may not represent real edges, but some noise or clutter. At higher scales, given by larger k values, $E(U^k)$ maps do not contain false edges, but, since they correspond to coarser details, some real boundaries could be missed or dislocated. Therefore, our edge tracking scheme starts at the finest scale of S and executes until it reaches the coarsest one. The boundaries that disappear fast when the scale is increased must be discarded, while those that exist at both fine and coarse scales should be kept, since they constitute salient structures. This fine-to-coarse tracking scheme considers all edge pixels of $E(U^1)$ and track them through the scale-space. If they appear in a large enough number of successive edge images $E(U^k)$, then can be considered as belonging to real edges and labeled accordingly, otherwise they are discarded. The tracking process goes on the same way, until there are not enough remaining scales so that the unlabeled edges of the current $E(U^k)$ to appear in a large enough number of them. Our choice for that number is $\lceil \frac{K}{2} \rceil$. The labeling process updates a binary matrix, initialized as $E(S) := 0$ and representing the detection result. The proposed tracking is formulated as [17]:

$$\forall k \in \left\{1, \ldots, \left\lfloor \frac{K}{2} \right\rfloor\right\}, i \in \{1, \ldots, I\}, j \in \{1, \ldots, J\} : E(U^k)[i, j]$$
$$= \cdots = E(U^{k+m})[i, j] = 1 \;\&\; m \geq \left\lceil \frac{K}{2} \right\rceil \Rightarrow E(S)[i, j] = 1 \tag{5.8}$$

This fine-to-coarse tracking algorithm can be improved by performing the search in small edge neighborhoods at higher scales. In order to address any possible edge displacement at large scales, an edge pixel would be accepted as valid even if it is not found at the same location in the edge map of the next scale, but in a neighborhood of it. The proposed improvement to be incorporated in (5.8) is formulated as:

$$\exists a \in \{i-1, i, i+1\}, b \in \{j-1, j, j+1\} : E(U^k)[i, j]$$
$$= E(U^{k+1})[a, b] = 1, \quad \forall k \geq high\,threshold \tag{5.9}$$

An edge detection example is displayed in Fig. 5.1. The edges of the [256 × 256] *Trui* image displayed in (a), which are detected at 10 scales of S are displayed in (b)–(k) and the final detection result is described in (l) [17].

This anisotropic diffusion-based edge detection framework was successfully tested on hundreds grayscale images from well-known image collections, such as

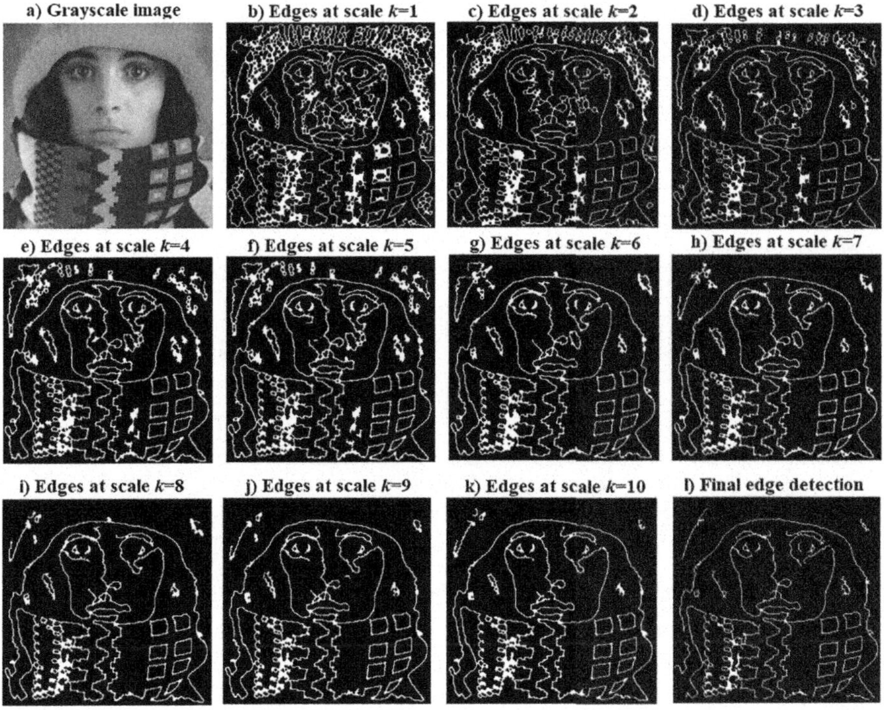

Fig. 5.1 Edge detection using a scale-space representation with $K = 10$ scales

the USC-SIPI Image Database and the Berkeley Segmentation Dataset (BSD) [17]. Optimal boundary detection results were obtained for $K = 15$ and $\tau = 40$.

The method comparison results show that our multi-scale edge extraction approach outperforms the classic detectors, achieving better performance measure scores. It is slightly outperformed by the state-of-the-art Canny filter that achieves somewhat better values of these metrics. An edge detection method comparison example is described in Fig. 5.2. Next, the average Performance Ratio (PR) and PSNR scores obtained by this technique and other detection filters are displayed in Table 5.1.

5.2 Energy and PDE-Based Image Segmentation Using Active Contours

Image segmentation represents an essential image analysis task whose purpose is to achieve a meaningful image partition. Each of the obtained segments has the property that its pixels are similar with respect to some property, such as color, intensity or texture [19]. The segmentation field has many important computer vision

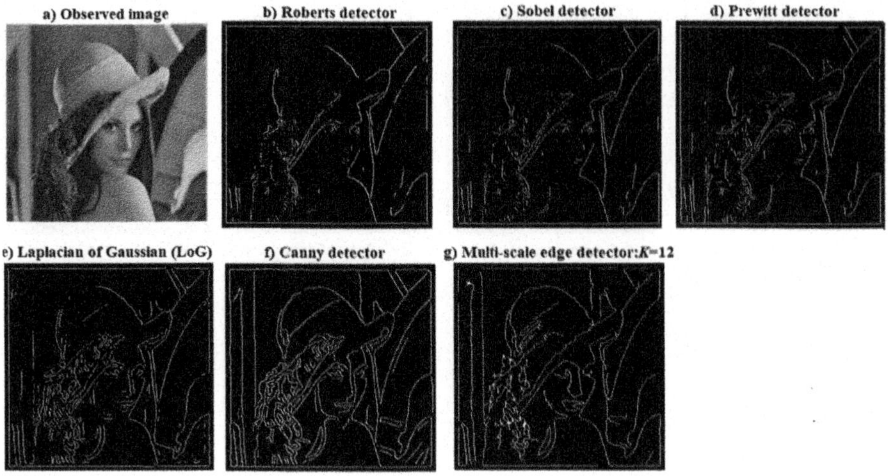

Fig. 5.2 An edge detection method comparison example

Table 5.1 Method comparison: average PR and PSNR values

Edge detection technique	Average PR	Average PSNR (in dB)
The proposed edge detector	11.0217	20.1742
Canny filter	12.1304	22.3817
Laplacian of Gaussian (Log)	9.3267	19.4837
Roberts operator	6.5486	15.6745
Sobel detector	7.4251	16.7802
Prewitt filter	8.5139	16.9837

application areas, including object detection and tracking, image object recognition, biometric authentication, image and video indexing and retrieval, video surveillance and medical imaging.

The existing image segmentation techniques methods could be grouped in: thresholding-based schemes, edge-based algorithms [20], histogram-based techniques [19], pixel clustering approaches [21], motion-based schemes [22], compression-based techniques [23], graph partitioning algorithms [24] and nonlinear PDE-based models. Active Contours, also called *snakes*, represent the most popular PDE-based segmentation models. They are based on curve propagation and have a variational character, since they perform energy-based functional minimizations. They are divided into two main categories: parametric and geometric models.

5.2.1 Parametric Active Contour Models

Introduced by Kass et al. in 1988 [25], the parametric active contours represent deformable models defined explicitly as parameterized curves. These snakes are deformable splines influenced by constraint and image forces that pull them towards the objects' contours and internal forces that resist deformation. A parametric active contour is simply a set of contour points $v(s) = (X(s), Y(s))$ parameterized by $s \in [0,1]$ that minimizes an energy associated to the contour as a sum of the internal and external energy: $E = E_{\text{internal}} + E_{external}$, where

$$E_{internal} = \frac{1}{2}(\alpha(s)|v_s(s)|)^2 + (\beta(s)|v_{ss}(s)|)^2$$
$$= \frac{1}{2}\left(\alpha(s)\left\|\frac{d\vec{v}}{ds}(s)\right\|\right)^2 + \left(\beta(s)\left\|\frac{d^2\vec{v}}{ds^2}(s)\right\|\right)^2 \quad (5.10)$$

$\alpha(s)$ and $\beta(s)$ represent some properly selected non-negative weights, and

$$E_{external} = -\int_0^1 f(X(s), X(s))ds, \quad \text{where} f(x, y) = |\nabla I(x, y)|^2 \quad (5.11)$$

where I is the analyzed image. The minimization of the contour energy E is solved applying two independent Euler equations that are next solved using the finite difference method [25].

This parametric active contour has obvious drawbacks, like the poor convergence performance for concave boundaries. Also, a higher accuracy of it requires tighter convergence criteria used for energy minimization and thus, a higher computation time. An object segmentation example based on this snake is described in Fig. 5.3 [25].

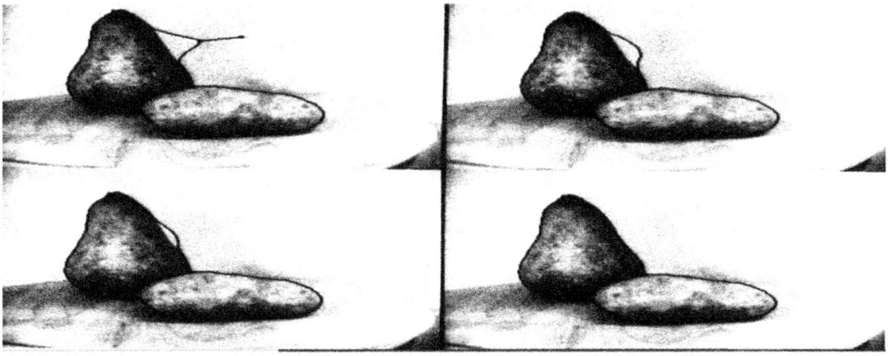

Fig. 5.3 Active contour evolving to the boundaries of some image objects

Some active contours that improve this snake model were developed in the last decades. An active contour that addresses the issues of the previous one is Gradient vector flow (GVF) snake model introduced by Xu and Prince in [26]. The *gradient vector flow* (GVF) field $\boldsymbol{v}$ minimizes the next energy functional:

$$E_{GVF} = \iint \mu\left(u_x^2 + u_y^2 + v_x^2 + v_y^2\right) + |\nabla f|^2 |v - \nabla f|^2 dxdy \quad (5.12)$$

where $f(x, y)$ is an edge map derived from $I(x, y)$. This energy minimization problem is solved applying 2 Euler equations, which are then solved through iteration towards a steady-state value [26].

This GVF-based active contour solves the poor convergence performance of the first snake. So, the example displayed in the next figure shows that an active contour with GVF external forces moves successfully into a concave boundary region [26] (Fig. 5.4).

The *Baloon models* represent active contours that have been developed so that to overcome the limitations of the default snake [27]. While the original snake is not attracted to distant boundaries and shrinks inwards if no substantial images forces are acting on it, the baloon model makes the curve behave like a balloon which is inflated by an additional force.

Its initial contour need no longer be close to the solution to converge, since it passes over the weak edges and stops only if the edge is strong enough. This active contour introduces the inflation term $F_{\text{inf}} = k_1\vec{n}(s)$, where k_1 is the force magnitude and $n(s)$ is normal unitary vector of the evolving curve, into the forces that act on

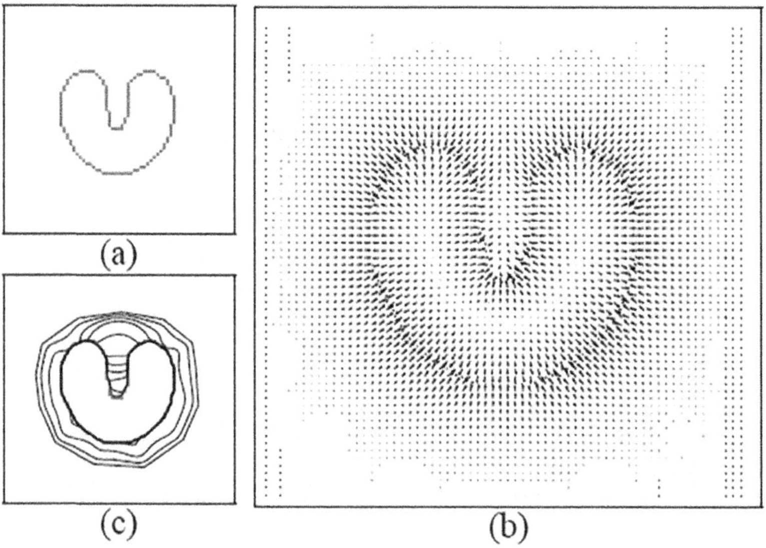

Fig. 5.4 A GVF snake evolving into a concave boundary region

the snake [27]. The baloon snake has its own drawbacks, such as the outward force making the contour slightly larger than the actual minima and the inflation force that may overpower forces from the weak edges.

Another category of deformable models is represented by the region-based active contours. The most popular of them is the piecewise-smooth Mumford and Shah model [28] and it is based on the next energy functional minimization:

$$E(u, C) = \frac{1}{2}\int_{\Omega} (f-u)^2 dx + \lambda^2 \frac{1}{2}\int_{\Omega - C} |\nabla u|^2 dx + v\| C\| \tag{5.13}$$

where f is the function of image to be segmented and C is the evolving contour.

This PDE variational model is ill posed and difficult to solve. Many solutions that aim minimize its energy functional have been proposed [29]. Also, other active contours modifying this Mumford–Shah energy functional have been developed.

Such an active contour is the *diffusion snake model*, introduced by Cremers et al. in 2001 [30]. It addresses several issues of the deformable models, such as their sensitiviy to noise, clutter, and occlusions. It defines the next energy for the contour C:

$$E(u, C) = E_i(u, C) + \alpha E_c(C) \tag{5.14}$$

where Ec favors contours familiar from a learning process and the other energy term represents the modified Mumford–Shah energy functional:

$$E_i(u, C) = \frac{1}{2}\int_{\Omega} (f-u)^2 dx + \frac{1}{2}\lambda^2 \int_{\Omega - C} |\nabla u|^2 dx + v\mathrm{L}(C) \tag{5.15}$$

with $\mathrm{L}(C) = \int_0^1 C_s^2 ds$ [30].

Thus, it represents a hybrid active contour combining the external energy of the Mumford–Shah energy functional with the Kass snake's internal energy. Next, the total energy E is minimized iteratively with respect to contour C and the segmented image u [30]. The corresponding Euler–Lagrange equation is determined and it is then solved applying the gradient descent method. It leads to an evolution equation which is approximated numerically. An image segmentation example based on the diffusion snake is described in Fig. 5.5 displaying the segmentation results with and without prior knowledge about object shape [30]. Other variants of this diffusion-based active contour model have been also proposed. Such a PDE-based segmentation model is the *simplified diffusion snake* [31].

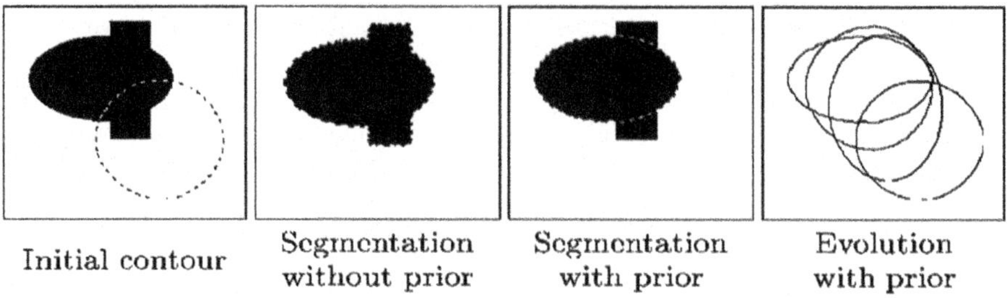

Fig. 5.5 Image segmentation applying the diffusion-based snake

5.2.2 Geodesic Active Contours

The *geometric*, or *geodesic, active contours* (GAC) represent deformable models that represent the contours implicitly as *level sets* of some 2D scalar functions. These segmentation models introduced by Caselles et al. in 1993, are active contours evolving in time according to the intrinsic geometric measures of the considered image [32].

The GAC models are achieved combining parametric active contours to level-sets and represent an effective tool to solve the limitations of the previously described parametric snakes. The *level set* of a real-valued function f is a set where the function takes on a given constant value c:

$$L_c(f) = \{(x_1, \ldots, x_n) | f(x_1, \ldots, x_n) = c\} \tag{5.16}$$

It becomes a level curve for $n = 2$, a level surface for $n = 3$ and a level hypersurface for $n > 3$. Also, if f is differentiable, the level set represents a hypersurface and a manifold outside this function's critical points.

The level sets have been successfully applied for both the contour- and region-based segmentation tasks. A geodesic active contour is based on energy minimization $\min \int_0^1 g(|\nabla I(C(q))|)\left|C'(q)\right|dq$ with C a parametrized planar curve [32]. One applies the Euler–Lagrange PDE that corresponds to it and gets the gradient descent curve evolution equation of GAC:

$$\frac{\partial C(t)}{\partial t} = g(I)\kappa\vec{\mathcal{N}} - \left(\nabla g \cdot \vec{\mathcal{N}}\right)\vec{\mathcal{N}} \tag{5.17}$$

where κ represents the Euclidean curvature and $\vec{\mathcal{N}}$ is the unit inward normal [32]. If C represents the level-set of the function u, it leads to the next evolution equation:

$$\frac{\partial u}{\partial t} = |\nabla u| div\left(g(I)\frac{\nabla u}{|\nabla u|}\right) + \nu g(I)\nabla u \tag{5.18}$$

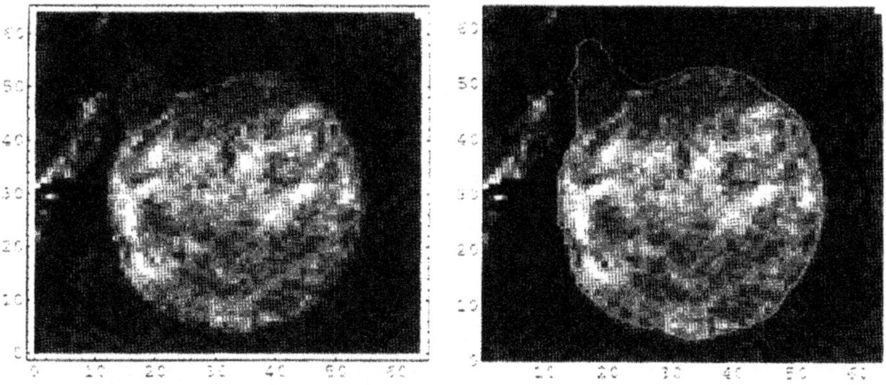

Fig. 5.6 A GAC-based tumor detection example

Since these geodesic flows have been applied successfully in the medical imaging field, a GAC-based medical image segmentation example depicting a tumor detection is described in Fig. 5.6 [32].

Another category of level-set based geometric active contours contains the *active contours without edges*. One of these segmentation models is that introduced by Chan and Vese in 1999 [33], which does not rely on the image boundaries at all.

This active contour, influenced by a simplified Mumford–Shah segmentation model, represents the segmentation boundary using a level set function [33]. If f is the image to be segmented, this variational model performs the energy minimization $\arg\min_{u,C} \mu\, \text{length}(C) + \int_\Omega (f(x) - u(x))^2 dx$ with $u(x) = \begin{cases} c_1, \textit{for x inside C} \\ c_2, \textit{for x outside C} \end{cases}$. It leads to the energy functional minimization:

$$\arg\min_{u,C} \mu\, \text{length}(C) + \nu\, Area(inside(C)) + \lambda_1 \int\limits_{inside(C)} |f(x) - c_1|^2 dx$$
$$+ \lambda_2 \int\limits_{outside(C)} |f(x) - c_2|^2 dx \tag{5.19}$$

Since C is represented as the zero-crossing of a level set function, $C = \{x \in \Omega : \varphi(x) = 0\}$, the energy minimization is rewritten as:

$$\arg\min_{c_1, c_2, \varphi} \mu \int\limits_\Omega \delta(\varphi(x)) |\varphi(x)| dx + \nu \int\limits_\Omega H(\varphi(x)) dx$$
$$+ \lambda_1 \int\limits_\Omega |f(x) - c_1|^2 H(\varphi(x)) dx + \lambda_2 \int\limits_\Omega |f(x) - c_2|^2 (1 - H(\varphi(x))) dx \tag{5.20}$$

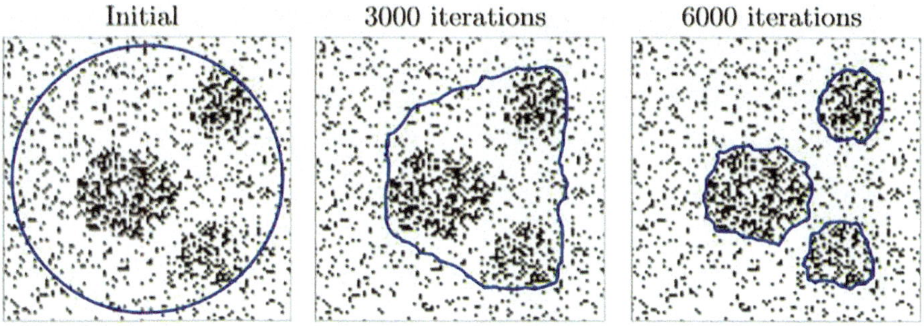

Fig. 5.7 Cluster detection using the Chan–Vese algorithm

where H is the Heaviside function and the function δ represents the Dirac measure [33].

$$c_1 = \frac{\int_\Omega f(x)H(\varphi(x))dx}{\int_\Omega H(\varphi(x))dx}, \quad c_2 = \frac{\int_\Omega f(x)(1 - H(\varphi(x)))dx}{\int_\Omega (1 - H(\varphi(x)))dx} \tag{5.21}$$

The variational Chan–Vese model is solved applying a semi-implicit gradient descent method [33]. A segmentation example related to this model is displayed in Fig. 5.7, where a cluster detection is performed successfully after 6000 iterations of this algorithm.

We also developed a level-set based variational segmentation inspired by the influential Chan–Vese algorithm, which is disseminated in [34]. Our approach performs the next energy functional minimization:

$$\varphi^* = \arg\min\{F(vf); \varphi \in X(\Omega)\} \tag{5.22}$$

where

$$\begin{aligned} F(\varphi) = {} & \nu \int_\Omega H(\varphi)dxdy + \mu \int_\Omega |\nabla H(\varphi)|dxdy \\ & + \lambda_1 \int_\Omega |u_0 - C_1(\varphi)|^2 H(\varphi(x))dxdy \\ & + \lambda_2 \int_\Omega |u_0 - C_2(\varphi)|^2 (1 - H(\varphi(x)))dxdy + \lambda_3 \int_\Omega |\varphi|^2 dxdy \end{aligned} \tag{5.23}$$

Next, the nonlinear PDE-based variational scheme in (5.17) was solved numerically by applying a convergent numerical approximation algorithm that was constructed using the gradient-descent scheme. That numerical solver developed by us is detailed in [34]. It was applied successfully in our contour detection experiments on various image datasets, and satisfactory segmentation results were achieved.

The proposed PDE varational contour tracking model was also rigorously investigated. A robust mathematical justification of the considered level-set model was performed in [34].

Other effective PDE-based active contour models with level-set functions, proposed by us and inspired by the GACs in (5.17)–(5.19), were diseminated in [35, 36]. These image segmentation solutions are used in combination with some AI and CV-based models for video object detection and tracking, being introduced to improve those detection and tracking solutions.

Thus, we introduced a moving vehicle recognition and tracking framework in [35]. It combines a Gaussian Mixture Model (GMM)-based video foreground extraction to a level-set based frame segmentation in the vehicle detection stage. A motion-based vehicle tracking process is next performed on the obtained detections by applying a Kalman filtering model. Then, the detected vehicles are recognized by their type applying a convolutional neural network (CNN)-based classifier and the tracking results [35].

The GMM-based detection process that is described in [37] may generate vehicle bounding boxes that do not fit exactly the detected vehicles. Some of them may contain even more than one vehicle, since a moving region detected by GMMs in a crowded traffic video could have more adjacent vehicles.

This situation is illustrated in Fig. 5.8, which describes a GMM-based traffic frame foreground detection process. One of the video objects detected in (a) contains two moving vehicles. Therefore, an active contour-based frame segmentation technique that improves the GMM-based vehicle detection was also proposed in [35].

Thus, a GAC-based segmentation is performed within each vehicle sub-image, a level-set function-based active contours being toward the vehicle boundaries [35]. The proposed approach applies the following nonlinear diffusion-based model:

$$\begin{cases} \frac{\partial u}{\partial t} - \lambda\psi\left(\left\|\nabla^2 u\right\|\right)\nabla \cdot \left(\varphi(|\nabla v_\sigma|)\psi(\|\nabla u\|)\nabla u\right) - \alpha\|\nabla u\|\varphi(|\nabla v|) = 0, \\ u(0, x, y) = u_0(x, y), \quad \forall(x,y) \in \Omega \\ \frac{\partial}{\partial n}(t, x, y) = 0, \quad \forall(x,y) \in \partial\Omega, \quad \text{in}(0,T) \times \Omega \end{cases} \tag{5.24}$$

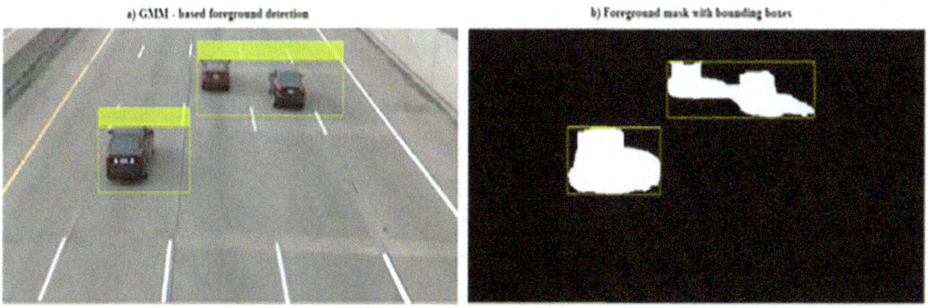

Fig. 5.8 GMM-based foreground detection for a traffic video frame

where u represents the level-set function, v is the vehicle-related image function, $\lambda, \alpha \in [0.6, 3)$, $\Omega \subseteq R^2$, $v_\sigma = v * G_\sigma$ and $G_\sigma(x, y) = \frac{1}{2\pi\sigma^2} e^{-\frac{x^2+y^2}{2\sigma^2}}$. The stopping function of this model is

$$\varphi : [0, \infty) \to [0, \infty) : \varphi(s) = e^{-s^2/\eta} \tag{5.25}$$

with $\eta \geq 5$, while the considered diffusivity function has the form:

$$\psi : [0, \infty) \to [0, \infty) : \psi(s) = \delta \sqrt[3]{\frac{\kappa}{|\xi + \beta s^2|}} \tag{5.26}$$

where $\beta \in (0, 1]$, $\delta, \xi, \kappa \in (1, 5]$.

Next, this nonlinear PDE-based model is solved numerically applying a finite difference method-based approximation algorithm that is detailed in [35]. The following explicit iterative numerical discretization scheme is thus determined for (5.24):

$$\begin{aligned} u_{i,j}^{n+1} &= u_{i,j}^n + \lambda\psi\left(u_{i+1,j}^n + u_{i-1,j}^n + u_{i,j+1}^n + u_{i,j-1}^n - 4u_{i,j}\right)\tau \\ &\sum_{q \in N(p)} \varphi\left(\left\|\nabla(v * G_\sigma)_{i,j}\right\|\right)\psi\left(\left\|\nabla u_{p,q}^n\right\|\right)\nabla u_{p,q}^n + \alpha\left\|\nabla u_{i,j}\right\|\varphi\left(\left\|\nabla v_{i,j}\right\|\right) \end{aligned} \tag{5.27}$$

where $n \in \{0, \ldots, N\}$ and $x = ih, y = jh, i \in \{1, \ldots, I\}, j \in \{1, \ldots, J\}$, for a $[Ih \times Jh]$ support image.

The numerical algorithm (5.27) is consistent to the proposed nonlinear diffusion-based model and converges quite fast to an optimal segmentation of the vehicle subimage, given by u^{N+1}, that represents a foreground mask which can be improved by applying some morphological operations. So, the low-sized connected components are removed from this binary image and the remaining ones correspond to the real vehicles [35].

A segmentation example using the proposed geometric active contour is provided in Fig. 5.9. Here, the 2 vehicles contained by the same moving object in Fig. 5.8a are separated by evolving the active contour in (5.27), with 60 iterations. The states u^0, u^{15}, u^{30}, u^{45}, u^{60} of the level-set function are depicted in the images (a)–(e) and the finally detected vehicles are provided in (f). As illustrated by these results, our active contour-based segmentation technique leads to much more reliable bounding boxes for the detected objects.

Another GAC-based segmentation approach that improve the vehicle detection results is proposed in [36]. There we introduced a CV-based multiple vehicle detection and tracking framework that applies a boosting algorithm for the detection task.

The proposed Haar Cascade Classifier based on the Gentle AdaBoost (GAD) meta-algorithm and trained in OpenCV by using some positive and negative vehicle

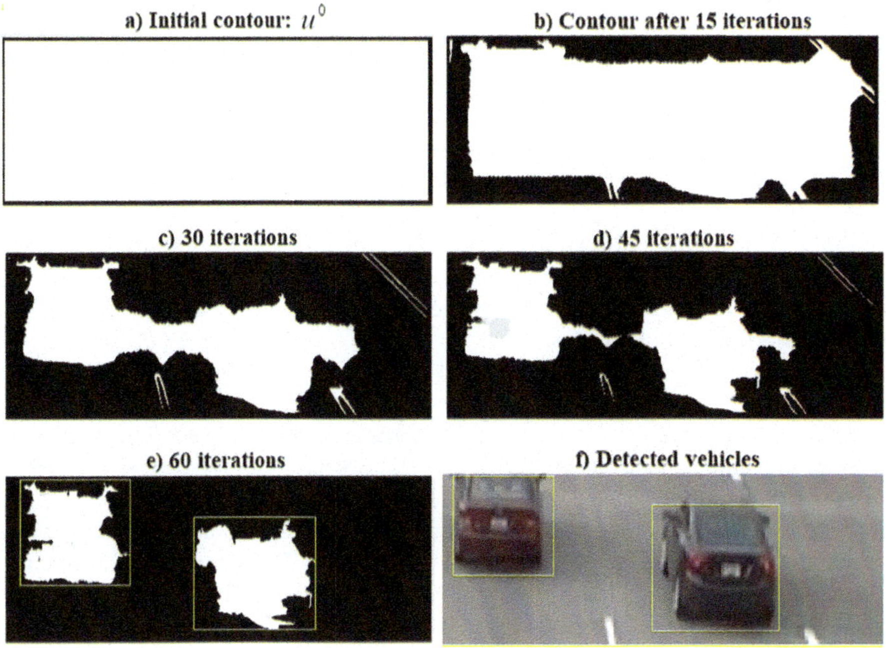

Fig. 5.9 Vehicle detection improved by active contour-based segmentation

datasets constructed by us, can also generate some vehicle bounding boxes that are much larger than the real vehicles and even could contain more than one vehicle [36].

So, the cascade classification-based vehicle detection results are improved by applying a PDE-based active contour segmentation within each of the subimages determined by those bounding boxes. The proposed GAC is based on the following nonlinear parabolic PDE model:

$$\begin{cases} \frac{\partial u}{\partial t} - \alpha\varphi\left(\frac{\|\Delta u\|+\|\nabla u\|}{2}\right) div(\psi(|\nabla v_\sigma|)\varphi(\|\nabla u\|)\nabla u) - \beta\psi(|\nabla v|)\|\nabla u\| = 0, \\ u(0, x, y) = u_0(x, y), \quad \forall(x,y) \in \Omega \\ \frac{\partial u}{\partial \vec{n}}(t, x, y) = 0, \quad \forall(x,y) \in \partial\Omega, \quad \text{in}(0,T) \times \Omega \end{cases} \tag{5.28}$$

where u is the level-set function of the evolving contour, v represents the function of the *vehicle* subimage, $\alpha, \beta \in [0.5, 2)$, the considered stopping function has the form:

$$\psi : [0, \infty) \to [0, \infty) : \psi(s) = e^{-s^3/\gamma} \tag{5.29}$$

with $\gamma \geq 4$, and the proposed diffusivity function is

$$\varphi : [0, \infty) \rightarrow [0, \infty) : \varphi(s) = \zeta \left(\frac{\lambda}{|\xi s^{\kappa} + \eta|} \right)^{\frac{1}{\kappa+1}} \tag{5.30}$$

where $\xi \in (0, 1]$ and $\lambda, \zeta, \kappa, \eta \in (1, 4]$. The intervals of all these parameters were determined empirically by applying the trial and error approach [36].

This nonlinear anisotropic diffusion-based segmentation model can be approximated numerically applying the finite difference method [36, 38]. The discrete observation u_0 is the initial contour represented as a binary mask that is almost equal in size to the subimage of v. This active contour converges rapidly to an optimal segmentation of it. It evolves u toward the boundaries of the vehicle(s) in that image, reaching them in less than 150 iterations.

As in the previous case, the resulted foreground mask is improved applying morphological procedures. Thus, one performs an opening followed by a closing operation, then applies an image filling process and removes the small-sized connected components [36].

A vehicle example based on the proposed technique is described in Fig. 5.10. The casade classifier-based detections in (a) are improved by the active contour-based segmentation, the final output being presented in (b). The segmentation of central bounding box, which contains two vehicles, is provided in (c), which displays the mask at several iteration moments [36]. The improved vehicle detection results show that our GAC-based segmentation approach generates bounding boxes that fit more correctly the vehicle objects.

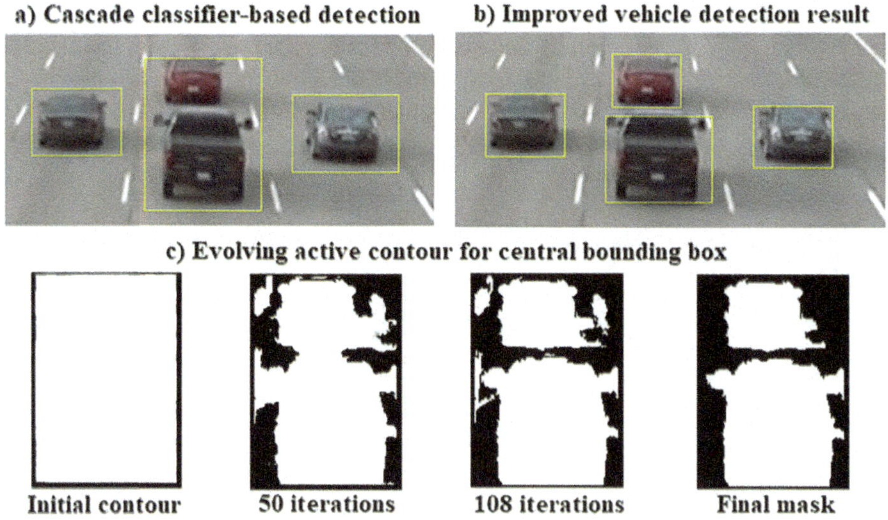

Fig. 5.10 Vehicle detection using cascade classification and active contours

5.3 Temporal Video Segmentation via PDE-Based Multiscale Analysis

Temporal video segmentation represents a significant computer vision domain [37]. This type of video image segmentation consists in dividing the video sequence in temporal segments, such as the video shots and scenes. It has a large variety of application fields, including the video compression, video database indexing and retrieval, video understanding and the video object detection and tracking.

While a videoshot represents a continuous serie of frames that are shot uninterruptedly by a single camera, a video scene constitutes a succession of semantically-correlated shots [39]. Video transitions, represent the mechanism used to change from one shot to the next one in a film sequence, can be categorized as follows: hard cuts, soft cuts and digital effects [37, 39, 40].

The hard cuts, also called simply cuts, represent the most encountered type of video transitions and constitute abrupt transitions between succesive shots. A soft cut is a gradual transition between two consecutive shots, representing a frame sequence belonging to both of them. The most common soft cuts represent fades or dissolves. The digital effects used as video shot transitions include the animated effects, wipes, lighting effects, color replacement, pixelization, focus drops and other effects.

Here one considers only the hard cut detection field. The video cut detection algorithms developed in the last decades could be grouped in several major classes. The first one contains the pixel difference-based cut detection techniques that measure the discontinuity of the visual content comparing the corresponding pixel intensities between two adjacent video frames. The sum of absolute differences (SAD) [37, 39, 40] and the pair-wise pixel comparisons [41] are used by these algorithms. The second category is that of the histogram comparison-based video cut detection methods which measure the similarity of the grayscale and color histograms corresponding to successive frames by applying metrics like the histogram difference, histogram quadratic distance and histogram intersection [39, 40].

The edge-based shot detection techniques are based on edge change ratio (ECR), edge tracking and edge histograms [37, 39, 40, 42]. The motion-based shot identification approaches apply motion estimation solutions to determine the motion breaks that could indicate the presence of some sudden transitions [42–44].

Statistical feature-based cut detection schemes break the frames into certain zones and compare various statistical measures, like those based on the average or standard deviation values, of the pixels in these regions [37, 39–41, 44]. Such a recently introduced fast statistical measure-based shot detection method uses the separable moments and SVM classifiers [45].

Other temporal segmentation solutions perform a simplification of the video image into a static one by applying the visual rhythm concept [46]. Some fuzzy logic-based shot boundary extraction techniques that can detect properly both the suddden and gradual video cuts were also developed [47]. Other shot transition detection algorithms combine the Principal Component Analysis (PCA) to deep learning models

[48, 49], while some cut detection techniques are based on multiple invariant features, including the edge change ratio, color descriptors and SIFT features [50].

We also introduced several effective temporal video segmentation techniques that overcome some drawbacks of the mentioned shot detections approaches. Some of those schemes, such as thoses using pixel differences, pair-wise pixel comparisons or edges, have an object and camera motion-sensitive characted and generate numerous false positives. Other algorithms, like those based on histograms, could provide many false negatives, because they disregard the spatial distribution. The detection solutions based on motion estimation and statistics are characterized by high computational costs and execution time. And many other existing techniques are not automatic or rely on error-generating thresholds.

One of the techniques developed by us applies a 2D Gabor filtering-based frame feature extraction, followed by an automatic inter-frame distance value clustering [51]. Another successful video segmentation approach developed by us is based on a high-level deep learning-based feature extraction [52]. The most recent temporal segmentation framework proposed by us is based on a successful combination of nonlinear second-order PDE models and convolutional neural networks [53].

The deep learning-based framework disseminated in [53] performs a high-level multi-scale feature extraction on the video frames, using a PDE-based scale-space representation. It is also fully automatic, since it does not use thresholding procedures and it is not motion-sensitive.

So, the following nonlinear second-order hyperbolic PDE-based model with boundary conditions was introduced to create the multiscale space:

$$\begin{cases} \alpha \frac{\partial^2 u}{\partial t^2} + \beta \frac{\partial u}{\partial t} - \eta \nabla \cdot (\psi(\|\nabla u_\sigma\|) \nabla u) + \lambda(u - u_0) = 0 \\ u(x, y, 0) = u_0(x, y), \quad \forall (x,y) \in \Omega \subseteq R^2 \\ u_t(x, y, 0) = u_1(x, y), \quad \forall (x,y) \in \Omega \\ u(x, y, t) = 0, \quad \forall (x,y) \in \partial\Omega \\ \frac{\partial u}{\partial n}(x, y, t) = 0, \quad \forall (x,y) \in \partial\Omega \end{cases}, \tag{5.31}$$

where $\alpha, \beta, \eta, \lambda \in (0, 1]$, $\partial\Omega$ is the frontier of the domain $\Omega \subseteq R^2$, the observed image $u_0 \in L^2(\Omega)$. The next positive, monotonically decreasing and convergent diffusivity function was considered [53]:

$$\psi : [0, \infty) \to [0, \infty), \psi(s) = \left(\frac{\xi}{|\delta s^k + \gamma|} \right)^{\frac{1}{k-1}}, \tag{5.32}$$

with $\delta \in (0, 1]$, $\xi \geq 4$, $\gamma \geq 5$ and $k \in \{2, 3, 4, 5\}$ [54].

The nonlinear PDE-based model (5.31) provides an effective detail-preserving image filtering and overcomes side-effects, like blurring and staircasing. Its second-time derivative, $\frac{\partial^2 u}{\partial t^2}$, that provides the hyperbolic character of this nonlinear PDE, sharpens the image boundaries. This PDE model is also non-variational and we demonstrated in [53] that it is well-posed under some certain conditions, its mathematical validity being rigorously treated.

Thus, we performed an integration operation on (5.31) and obtain the integral model [53]:

$$\begin{cases} \frac{\partial u}{\partial t}(x,y,t) - \frac{1}{\alpha}\int\limits_0^t e^{\frac{\beta^2}{\alpha}(t-s)}(\eta\, div(\psi(\|\nabla u_\sigma(x,y,s)\|)\nabla u(x,y,s)) + \lambda(u(x,y,s) - u_0(x,y)))ds = \\ u(0,x,y) = u_0(x,y) \\ u(x,y,t) = 0, \quad \text{on } \partial\Omega \times (0,T) \end{cases} \tag{5.33}$$

A modified variant of this integral model may be well-posed under some certain assumptions [53]. So, (5.33) can be replaced by the following model:

$$\begin{cases} \frac{\partial u}{\partial t} - \varphi\Delta u - \int\limits_0^t (\eta\, div(\psi(\|\nabla u_\sigma\|)\nabla u) + \lambda(u - u_0))ds = 0 \\ u(x,y,0) = u_0(x,y) \\ u(x,y,t) = 0 \quad \text{on } \partial\Omega \times (0,T) \end{cases} \tag{5.34}$$

This integral equation admits a solution for $\varphi > 0$ if these conditions hold [53]:

$$\begin{cases} (\psi(\|v\|)v - \psi(w)w)(v - w) \geq 0, \quad \forall v, w \in R^2 \\ \exists a, b : a \geq \psi(s) \geq b > 0, \quad \forall r \geq 0 \\ \psi - continuous \end{cases} \tag{5.35}$$

We have:

$$\frac{\partial(\psi(s)s)}{\partial s} = \psi'(s)s + \psi(s) \geq 0, \quad \forall s \in R^+, \tag{5.36}$$

since $\psi'(s) \geq 0$, which means function $s \to \psi(s)s$ is monotone in R^2, therefore the first condition of (5.35) holds. Its second condition also holds, since ψ is bounded. Because it is also continuous on $[0, \infty)$, its third condition also holds [53].

These three conditions being satisfied, the integral Eq. (5.34) admits a unique variational solution $u^* : \Omega \times (0,T) \to \mathbb{R}$ in sense of distributions [55], with $T > 0$. This means

$$u^* \in L^\infty\left(0, T; H_0^1(\Omega)\right), \frac{\partial u^*}{\partial t} \in L^2\left(0, T; L^2(\Omega)\right), \tag{5.37}$$

$$\nabla \cdot \left(\psi\left(\left\|\nabla u_\sigma^*\right\|\right)\nabla u^*\right) \in L^\infty\left(0, T; L^2(\Omega)\right) \tag{5.38}$$

and

$$\int_{\Omega} \frac{\partial}{\partial t} u^*(x, y, t)\chi(x, y)dxdy + \varphi \int_{\Omega} \nabla u^*(x, y, t) \cdot \nabla \chi(x, y)dxdy$$
$$+ \int_0^t ds \int_{\Omega} \eta\psi\left(\left\|\nabla u_\sigma^*(x, y, t)\right\|\right)\nabla u^*(x, y, t) \cdot \nabla \chi(x, y)$$
$$+ \lambda\left(u^*(x, y, s) - u_0(x, y)\right)\chi(x, y)dxdy = 0,$$
$$u^*(x, y, 0) = u_0(x, y), \quad \forall x, y \in \Omega, \quad \forall \chi \in H_0^1(\Omega) \tag{5.39}$$

where the Sobolev space $H_0^1(\Omega) = \left\{u \in L^2(\Omega); \nabla u \in \left(L^2(\Omega)\right)^2, u = 0 \text{ on } \partial\Omega\right\}$ [54].

The integral model (5.34) represents a very good approximation of (5.33) and, for $\varphi \to 0$, the solution of (5.34) converges in a certain weak sense to a solution of (5.33). This well-posed nonlinear second-order hyperbolic model is solved numerically by applying a finite difference-based numerical approximation algorithm that converges to its weak solution. The same a grid of space size h and the time step Δt, which has been used for the other discretization tasks, is applied here too. The construction of the numerical approximation scheme is detailed in [53]. Its final explicit form is the following one:

$$u_{i,j}^{n+1} = u_{i,j}^{n}\left(\frac{4\lambda - 2\alpha}{\beta + 2\alpha}\right) + u_{i,j}^{n-1}\left(\frac{\beta - 2\alpha}{\beta + 2\alpha}\right) + u_{i,j}^{0}\frac{2\lambda}{\beta + 2\alpha}$$
$$+ \eta\varsigma \sum_{q \in N_p} \psi\left(\left|\nabla(u * G_\sigma)_{p,q}^n\right|\right)\nabla u_{p,q}^n \tag{5.40}$$

where $n \in \{0, \ldots, N\}$. This explicit iterative numerical discretization scheme in (5.40) is stable and consistent to the hyperbolic PDE model and converges quite fast to its weak solution. It is used successfully to construct the scale-space representation for the multi-scale feature extraction [53].

The analyzed video is represented as $V = [F_1, \ldots, F_M]$, where F_i are the frames and $i \in \{1, \ldots, M\}$. The PDE-based scale-space is created applying the numerical scheme (5.40) on the current frame and considering the filtering results at several iteration moments. Each frame F_i is converted to the de-correlated color space CIE L*a*b* and the filtered L*a*b frame is $\left[(L(F_i))^n, a(F_i), b(F_i)\right]$, where $(L(F_i))^n$ represents the luminance channel filtering after n iterations, is converted back to RGB form. The obtained multi-scale representation of K scales of F_i is:

$$S(i) = \left\{F_i, RGB\left(\left[(L(F_i))^r, a(F_i), b(F_i)\right]\right), \ldots, RGB\left(\left[(L(F_i))^{r(K-1)}, a(F_i), b(F_i)\right]\right)\right\} \tag{5.41}$$

where $r \in [3, 10]$ is the iteration step and $K \geq 3$ [53].

High-level characteristics of the frames are extracted applying a deep learning-based approach. A DL-based feature extraction is applied on the current

F_i at each scale $k \in \{0, \dots, K-1\}$, where it has the form $S(i)\{k\} = RGB\big(\big[(L(F_i))^{rk}, a(F_i), b(F_i)\big]\big)$.

The trained CNNs, or ConvNets, can learn high-level feature representations for various types of images, generating powerful content descriptors [56]. We combined 2 deep neural networks to compute the high-level charateristics of each frame: Inception-ResNet-V2 and DenseNet-201 [54, 57].

The first network, Inception-ResNet-V2, is a DL-based model that builds on the Inception family while incorporating residual connections [54]. It has 164 layers was trained on more than 1 million images of ImageNet collection. The second one, DenseNet-201, with 201 layers, represents a ConvNet also trained on ImageNet and characterized by dense connections between layers, through Dense Blocks, where one connects all layers directly with each other, in a feed-forward way [57]. The architectures of these deep learning networks are detailed in [53].

The network activations are determined by forward propagating the input through each CNN up to the specified layer. While the layers of these CNNs generate activations on the input images, not all of them have the same of feature extraction capacity. So, the first convolutional layers could compute only low-level characteristics, which are then processed by the deeper layers that combine them such that to achieve higher level descriptors.

Our combined approach uses the fully connected layers of the networks: the *predictions* layer of Inception-ResNet-v2, located before final classification layer, and FC 1000, with 1000 neurons, of the DenseNet-201. Therefore, each $S(i)\{k\}$ of F_i is pre-processed according to the specifications of Inception-ResNet-V2's input layer: it is resized at $[299 \times 299 \times 3]$ and next 0-center normalization is applied:

$$S(i)\{k\} := \frac{S(i)\{k\} - \mu(S(i)\{k\})}{\sigma(S(i)\{k\})}, \quad \forall k \in \{0, \dots, K-1\} \tag{5.42}$$

Then, the processed $S(i)\{k\}$ is fed into the Inception-ResNet-V2 and its predictions layer generates an activation that determines the high-level features in the form of the feature vector $V_{IRN}(S(i)\{k\})$ with 1000 coefficients. The initial version of $S(i)\{k\}$ is also pre-processed according to the DenseNet-201 specifications. It is resized at the input size format of $[224 \times 224 \times 3]$ and the 0-center normalization provided by (5.42) is performed on it. The pre-processed image is fed into DenseNet-201 model, whose FC 1000 layer generates an activation producing the high-level feature vector $V_{DN}(S(i)\{k\})$, with 1000 coefficients too. A 2D feature vector is achieved at this scale by concatenating these 2 vectors as follows:

$$V_i(k) := [V_{IRN}(S(i)\{k\});\ \ V_{DN}(S(i)\{k\})] \tag{5.43}$$

The DL-based 2D feature vectors determined at multiple scales are combined into the final two-dimension descriptor of the frame, as follows:

$$Fv(i) := [V_i(0) \dots V_i(K-1)], \quad \forall i \in \{1, \dots, M\} \tag{5.44}$$

These $[2 \times 1000K]$ high-level feature vectors $Fv(i)$ are powerful content descriptor of the frames F_i. The high-level frame characterizations provided by them lead to successful discriminations between the frames of the video V [53].

These frames are then grouped in shots using these feature vectors. Other segmentation methods, which detect the cuts using threshold-based procedures, may generate numerous errors, since the optimal threshold selection is still a difficult task. For this reason, our temporal segmentation approach does not rely on any thresholds for shot break detection, applying a video frame clustering process instead of thresholding. That clustering procedure is completely automatic, since our shot detection technique uses no a-priori knowledge about the number of cuts [53].

Thus, one determines the set of the inter-frame difference metric values $\{d_1, \ldots, d_{M-1}\}$, where $d_i = d(Fv(i), Fv(i+1))$, $i \in \{1, \ldots, M-1\}$ and d is a metric working for these vectors, for example the 2D Euclidean metric. Any inter-shot distance value must be much higher than any intra-shot distance value and these distance values satisfy this condition [53]. So, we have:

$$\min_{i \in C} d_i \gg \max_{j \in \{1,\ldots,M-1\} \setminus C} d_j, \tag{5.45}$$

where $C \subseteq \{1, \ldots, M-1\}$ contains the indices of the cuts that must be identified.

The values of the set $\{d_1, \ldots, d_{M-1}\}$ are grouped into *high distances* and *low distances*. The high feature vector distance values correspond to the abrupt transitions in the video V. An automatic unsupervised classification algorithm based on agglomerative hierarchical clustering with an average-linkage clustering metric is applied on $\{d_i\}_{i \in \{1,\ldots,M-1\}}$ [58, 59]. Since this set must be divided into 2 classes of distance values, the number of clusters for the hierarchical clustering is 2. Our technique labels each distance value with either 1, for *low*, or 2, for *high*. Since $label(d_i) = 2 \Leftrightarrow i \in C$, the positions of the cuts in V are thus identified. The obtained set C determines the temporal segments of video V [53].

A temporal video image segmentation example based on the proposed approach is described in Fig. 5.11. The shot breaks of a video composed of 475 $[720 \times 1280 \times 3]$ frames were detected at the locations of the *high* feature vector distance values depicted in (*k*). The detected shot cuts are represented as pairs of frames in (a)–(j) [53].

This CNN-based multi-scale video segmentation framework was tested successfully on many videos [53]. It achieved a high detection rate, generating only few false positives and negatives, and high scores of the performance measures: *Precision, Recall* and F_1. Given its deep learning-based feature extraction component, this technique outperforms segmentation methods based on pixel differences, color, grayscale or edge histograms, statistics-based features, SIFT, LBP, SURF and 2D Gabor filter-based descriptors. Some method comparison results are presented in Table 5.2. Our framework achieves higher average *Precision, Recall* and F_1 values than the other techniques [53].

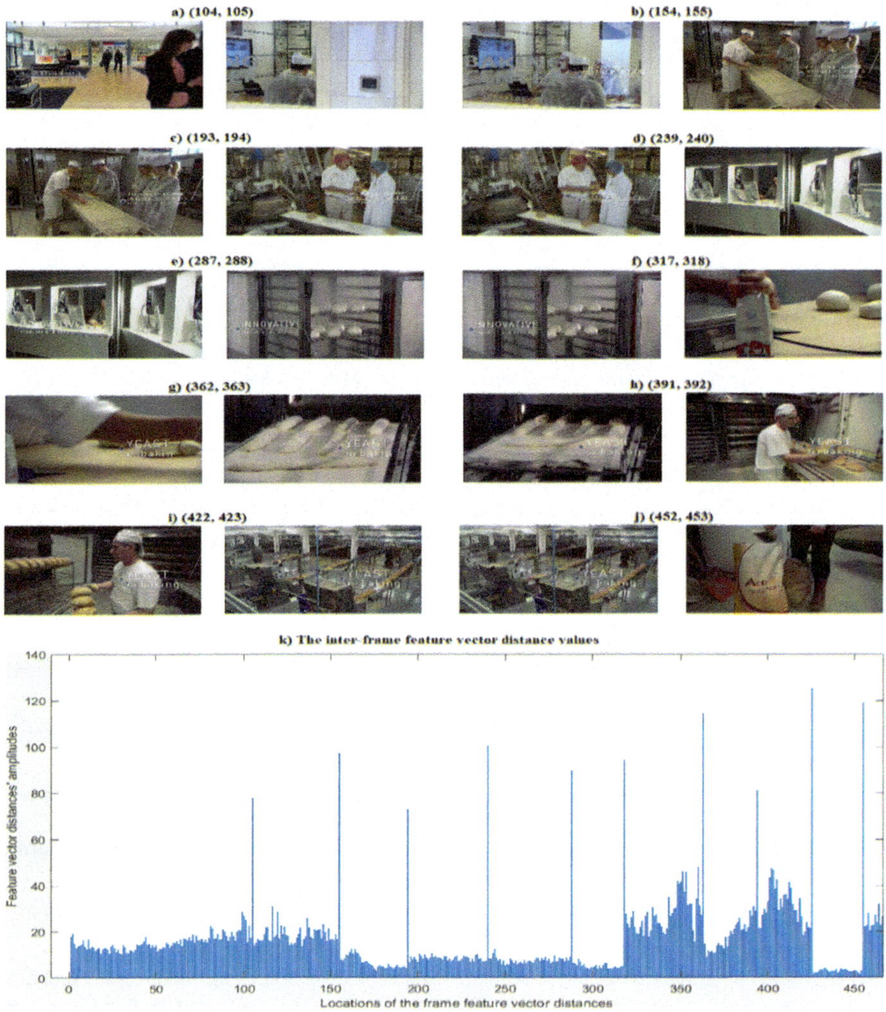

Fig. 5.11 Shot break detection using the proposed approach

Table 5.2 Video temporal segmentation method comparison results

Technique	Precision	Recall	F1
The proposed method	0.984	0.991	0.987
Color histograms	0.745	0.724	0.734
Edge histograms	0.815	0.752	0.782
Pixel differences (SAD)	0.775	0.763	0.769
Gabor 2D filter-based model	0.941	0.840	0.887
Statistical features with LHR	0.645	0.618	0.631
Pairwise pixel comparisons	0.741	0.723	0.731

Also, this approach properly in both clean and noisy image conditions. But, while it provides an effective hard cut detection, it achieves weaker segmentation results for other types of shot transitions, such as the gradual transitions and digital effects.

References

1. D. Ziou, S. Tabbonne, Edge detection techniques: An overview, *International Journal of Pattern Recognition and Image Analysis*, vol. 8, no. 4, pp. 537–559, 1998.
2. G. N. Chaple, R. D. Daruwala, M. S. Gofane, Comparisons of Robert, Prewitt, Sobel operator based edge detection methods for real time uses on FPGA, *International Conference on Technologies for Sustainable Development (ICTSD)*, Mumbai, India, pp. 1–4, 2015.
3. R. Haralick, Digital step edges from zero crossing of second directional derivatives, *IEEE Transactions on Pattern Analysis and Machine Intelligence*, vol. 6, no. 1, pp. 58–68, 1984.
4. D. Marr, E. Hildreth, Theory of Edge Detection, *Proceedings of the Royal Society of London*, Series B, Biological Sciences, **207**, no. 1167, pp. 187–217, 29 Feb 1980.
5. J. Canny, A computational approach to edge detection, *IEEE Transactions on Pattern Analysis and Machine Intelligence*, vol. 8, pp. 679–714, 1986.
6. X. Ren, Multi-scale improves boundary detection in natural images, *European conference on computer vision*, Springer, Berlin, Heidelberg, pp. 533–545, 2008.
7. A. P. Witkin, Scale-space filtering, *International Joint Conference on AI*, pp. 1019–1022, 1983.
8. X. Zhuang, T. Huang, Multi-scale Edge Detection with Gaussian Filters, ICASSP, *International Conference on Acoustics, Speech and Signal Processing-Proceedings. Institute of Electrical and Electronics Engineers Inc.*, pp. 2047–2050, 1986.
9. F. Bergholm, Edge focusing, *IEEE PAMI*, 9(6), pp. 726–741, 1987.
10. J. Williams, M. Shah, Edge contours using multiple scales, *Computer vision, graphics, and image Processing*, 51, 3, pp. 256–274, 1990.
11. Y. Li, Y. Bi, W. Zhang, C. Sun, Multi-scale anisotropic Gaussian kernels for image edge detection, *IEEE Access*, 8, pp. 1803–1812, 2019.
12. C. Lopez-Molina, B. De Baets, H. Bustince, J. Sanz, E. Barrenechea, Multiscale edge detection based on Gaussian smoothing and edge tracking, *Knowledge-Based Systems*, 44, pp. 101–111, 2013.
13. W. F. Ma, C. X. Deng, An improved wavelet multi-scale edge detection algorithm, *International Conference on Wavelet Analysis and Pattern Recognition*, IEEE, pp. 302–306, 2012.
14. W. Sun, J. Song, L. Zhang, Wavelet multi-scale edge detection using adaptive threshold, 5th *International Conference on Wireless Communications, Networking and Mobile Computing*, IEEE, pp. 1–4, 2009.
15. P. Perona, J. Malik, Scale-space and edge detection using anisotropic diffusion, *IEEE Transactions on pattern analysis and machine intelligence*, 12, 7, pp. 629–639, 1990.
16. T. Barbu, Segmentation-Based Non-Texture Image Compression Framework Using Anisotropic Diffusion Models, *Proceedings of the Romanian Academy, Series A*, Volume 20, Issue 2, pp. 122–130, April - June 2019.
17. T. Barbu, Automatic edge detection solution using anisotropic diffusion-based multi-scale image analysis and fine-to-coarse tracking, *Proc. Rom. Acad., Ser. A, Math. Phys. Tech. Sci. Inf. Sci*, 22, 267–274.
18. T. Y. Zhang, C. Y. Suen, A fast parallel algorithm for thinning digital patterns, *Commun. ACM* 27, pp. 236–239, 1984.
19. L. G. Shapiro, G. C. Stockman, *Computer Vision*, pp. 279–325, New Jersey, Prentice-Hall, 2001.
20. T. Lindeberg, M.-X. Li, Segmentation and classification of edges using minimum description length approximation and complementary junction cues, *Computer Vision and Image Understanding*, 67 (1), pp. 88–98, 1997.

21. L. Barghout, J. Sheynin, Real-world scene perception and perceptual organization: Lessons from Computer Vision, *Journal of Vision*, 13 (9): 709. https://doi.org/10.1167/13.9.709
22. J. Kenney, T. Buckley, O. Brock, Interactive segmentation for manipulation in unstructured environments, *IEEE International Conference on Robotics and Automation*, pp. 1377–1382, 2009, IEEE.
23. M. Hossein, R. Shankar, Y. Allen, S. Shankar, M. Yi, Segmentation of Natural Images by Texture and Boundary Compression, *International Journal of Computer Vision*, 95: pp. 86–98, 2011.
24. J. Shi, J. Malik, Normalized Cuts and Image Segmentation, *IEEE Transactions on Pattern Analysis and Machine Intelligence*, pp 888–905, Vol. 22, No. 8, 2000.
25. M. Kass, A. Witkin, D. Terzopoulos, Snakes: Active contour models, *International journal of computer vision*, 1(4), pp. 321–331, 1988.
26. C. Xu, J. L. Prince, Gradient vector flow: A new external force for snakes, *Proceedings of IEEE computer society conference on computer vision and pattern recognition*, pp. 66–71, 1997, IEEE.
27. L. D. Cohen, On active contour models and balloons, *CVGIP: Image Understanding*, Volume 53, Issue 2, pp. 211–218, March 1991.
28. D. Mumford, J. Shah, Optimal approximations by piecewise smooth functions and associated variational problems, *Communications on Pure and Applied Mathematics*, Vol. 42, No. 5, pp 577–685, 1989.
29. B. Bourdin, A. Chambolle, Implementation of an adaptive finite-element approximation of the Mumford-Shah functional, *Numerische Mathematik*, 85(4), pp. 609–646, 2000.
30. D. Cremers, C. Schnorr, J. Weickert, Diffusion-snakes: combining statistical shape knowledge and image information in a variational framework, *Proceedings IEEE Workshop on Variational and Level Set Methods in Computer Vision*, pp. 137–144, 2001.
31. D. Cremers, F. Tischhäuser, J. Weickert, C. Schnörr, Diffusion snakes: Introducing statistical shape knowledge into the Mumford-Shah functional, *International journal of computer vision*, 50 (3), pp. 295–313, 2002.
32. V. Caselles, F. Catté, T. Coll, F. Dibos, A geometric model for active contours in image processing, *Numerische mathematik*, 66 (1), pp. 1–31, 1993.
33. T. Chan, L. Vese, An active contour model without edges, *International Conference on Scale-Space Theories in Computer Vision*, pp. 141–151, Springer, Berlin, Heidelberg, 1999.
34. T. Barbu, Robust contour tracking model using a variational level-set algorithm. *Numerical Functional Analysis and Optimization*, 35 (3), pp. 263–274, 2014.
35. T. Barbu, S. I. Bejinariu, CNN-based Moving Vehicle Recognition using GMM-based Foreground Modeling, Level-set based Segmentation and Kalman Filter-based Tracking, *International Conference on INnovations in Intelligent SysTems and Applications (INISTA)*, pp. 1–6, Sept. 2024, IEEE.
36. T. Barbu, S. I. Bejinariu, R. Luca, Automatic Vehicle Detection and Tracking Using Cascade Classifiers and CNN-Based Multi-Scale Feature Extraction, *2024 IEEE Conference Advanced Topics on Measurement and Simulation (ATOMS)*, pp. 1–4, aug. 2024.
37. I. Koprinska, S. Carrato, Temporal video segmentation: a survey. *Signal Process: Image Communication*, 16 (5), pp. 477–500, 2001.
38. P. Johnson, *Finite Difference for PDEs*, School of Mathematics, University of Manchester, Semester I, 2008.
39. X. Zhu, W. G. Aref, J. Fan, A. C. Catlin, A. K. Elmagarmid, Medical video mining for efficient database indexing, management and access, *Proceedings of 19th IEEE international conference on data engineering (ICDE 2003)*, India, March, 2003.
40. J. S. Boreczky, L. A. Rowe, Comparison of video shot boundary detection techniques. *Journal of Electronic Imaging* 5 (2), pp. 122–128, April 1996.
41. U. Patel, P. Shah, P. Panchal, Shot detection using pixel wise difference with adaptive threshold and color histogram method in compressed and uncompressed video, *Int. J. Comput Appl.* 64 (4), pp. 38–44, 2013.

42. A. Jacobs, A. Miene, G. T. Ioannidis, O. Herzog, Automatic shot boundary detection combining color, edge, and motion features of adjacent frames, *TRECVID 2004 Workshop Notebook Papers*, pp. 197–206, 2004.
43. A. M. Tekalp, *Digital Video Processing*, Prentice-Hall, 1995.
44. Y. Yusoff, W. Christmas, J. Kittler, Video shot cut detection using adaptive thresholding, *Proceedings of the 11th British machine vision conference*, university of Bristol, September 11–14, 2000.
45. Z. N. Idan, S. H. Abdulhussain, B. M. Mahmmod, K. A. Al-Utaibi, S. A. R. Al-Hadad, S. M. Sait, Fast Shot Boundary Detection Based on Separable Moments and Support Vector Machine, *IEEE Access 2021*, vol. 9, pp. 106412–106427, 2021.
46. S. J. F. Guimarães, M. Couprie, Video segmentation based on 2d image analysis, *Pattern Recognition Letters*, 24 (7), pp. 947–957, 2003.
47. H. Fang, J. Jiang, Y. Feng, A fuzzy logic approach for detection of video shot boundaries, *Pattern Recognition*, 39 (11), pp. 2092–2100, 2006.
48. D. Chakraborty, W. Chiracharit, K. Chamnongthai, Video Shot Boundary Detection Using Principal Component Analysis (PCA) and Deep Learning, *18th International Conference on Electrical Engineering / Electronics, Computer, Telecommunications and Information Technology (ECTI-CON)*, 2021, pp. 272–275.
49. J. Xu, L. Song, R., Shot Boundary Detection Using Convolutional Neural Networks, *Visual Communications and Image Processing (VCIP), 2016.*
50. J. T. Jose, S. Rajkumar, M. R. Ghalib, A. Shankar, P. Sharma, M. R. Khosravi, Efficient Shot Boundary Detection with Multiple Visual Representations, *Mobile Information Systems*, vol. 2022, Article ID 4195905, 14 pages, 2022.
51. T. Barbu, Novel automatic video cut detection technique using Gabor filtering, *Computers & Electrical Engineering*, Volume 35, Issue 5, pp. 712–721, 2009.
52. T. Barbu, Automatic Deep Learning-based Temporal Video Segmentation Framework, *International Symposium on Signals, Circuits and Systems (ISSCS)*, pp. 1–4, 2021. IEEE.
53. T. Barbu, CNN-based temporal video segmentation using a nonlinear hyperbolic PDE-based multi-scale analysis, *Mathematics*, 11 (1), 245, MDPI, 2023.
54. C. Szegedy, S. Ioffe, V. Vanhoucke, A. Alemi, Inception-v4, Inception-ResNet and the Impact of Residual Connections on Learning, *AAAI*, vol. 4, p. 12, 2017.
55. V. Barbu, *Nonlinear Semigroups and Differential Equations in Banach Spaces*, Noordhoff International Publishing, 1976.
56. J. Murphy, *An overview of convolutional neural network architectures for deep learning*, Microway Inc, 2016.
57. H. Gao, Z. Liu, L. Van Der Maaten, K. Q. Weinberger, Densely Connected Convolutional Networks, *CVPR*, vol. 1, no. 2, p. 3. 2017.
58. T. Barbu, An automatic unsupervised pattern recognition approach, *Proceedings of the Romanian Academy, Series A*, January-April, Volume 7, Number 1, pp. 73–78, 2006.
59. T. Barbu, Unsupervised SIFT-based face recognition using an automatic hierarchical agglomerative clustering solution, *Procedia Computer Science*, Vol. 22, Elsevier, pp. 385–394, 2013.

Chapter 6
Nonlinear PDE-Based Video Object Detection and Tracking

The applications of the nonlinear PDE models in the video object detection and tracking field are discussed in this chapter. This computer vision domain has important application areas, including video monitoring, law enforcement, security systems, human–computer interaction, video indexing and retrieval, robotic vision, medical imaging and augmented reality. A detection and tracking task means locating some certain objects in the frames of a video sequence and identifying their trajectories. Besides the active contour models described in the previous chapter, video object detection use also methods based on template matching [1], frame differencing [2], dictionary-based models [3], deformable part-based models [4], cascade classifiers [5], invariant descriptors with SVM-based classifiers [6-8], Aggregate Channel Features (ACF) [9], GMM [10] and deep learning [11]. The video tracking approaches use Kalman filters [12], mean-shift schemes [13], Hidden Markov Models (HMM) [14], fragment-based models [15], Particle Swarm Optimization (PSO) [16] and CNN [17]. The active contours, video optical flow estimation and tracking-by-detection (TBD) methods that are presented here represent also useful tracking tools. Thus, the active contours based on variational PDE models, which are used successfully for solving the object detection tasks, can be also adapted for object tracking. These active contour-based video tracking techniques are described in the first section. Next, the detection and tracking of the moving objects in video sequences using the optical flow-based motion estimation is presented in the second section. Some variational PDE-based optical flow estimation models are described here. In the third section, the PDE-based video object tracking by detection solutions are discussed. Our own contributions in this computer vision area, representing tracking by detection techniques using nonlinear anisotropic diffusion-based multi-scale image analysis models, are introduced in this final section.

T. Barbu, *Digital Image Processing, Analysis and Computer Vision Using Nonlinear Partial Differential Equations*, Studies in Computational Intelligence 1211,
https://doi.org/10.1007/978-3-031-89576-0_6

6.1 Active Contour-Based Video Object Tracking Techniques

The nonlinear PDE-based active contours described in Chap. 5 represent effective image segmentation solutions that can be used successfully to solve the object detection tasks in both static and video images. But they could also be applied to video object detection and tracking, since the models presented in the previous chapter can be extended to detection in 3D images and video sequences. Thus, the contour tracking techniques usually evolve the object contour at the previous time (frame) towards its new location at the current time (frame).

The geodesic active contour model proposed by Caselles et al. in [18] is adapted successfully to moving object detection and tracking in [19]. It follows the boundaries of the moving objects in a video sequence of frames $I(t, x)$, where t is the continuous time, by using a modified GAC model. So, a term providing the object displacement between the video frames is inserted in Eq. (5.18), the following extended geometric active contour being obtained:

$$\frac{\partial u}{\partial t} = |\nabla u| div\left(g(t,x)\frac{\nabla u}{|\nabla u|}\right) + \nu g(t,x)|\nabla u| + (1 - g(t,x))v \cdot \nabla u \qquad (6.1)$$

where v represents the velocity vector field and g constitutes a smooth function [19].

The PDE model (6.1) moves the contours to the boundary of the moving objects. These object contours are then considered as the initial estimates of the contours in the next frame, and this contour tracking process continues iteratively. A contour tracking example based on the new GAC is described in Fig. 6.1 [19]. While the image in (a) shows the initial contour inside the moving ball, the images (c) and (d) illustrate the contour evolved according to Eq. (6.1).

Another well-known geodesic active contour-based video object detection and tracking framework is that introduced by Paragios and Deriche in [20]. Its detection component is based on the following minimization of the energy functional associated to the given curve $C(p, t)$:

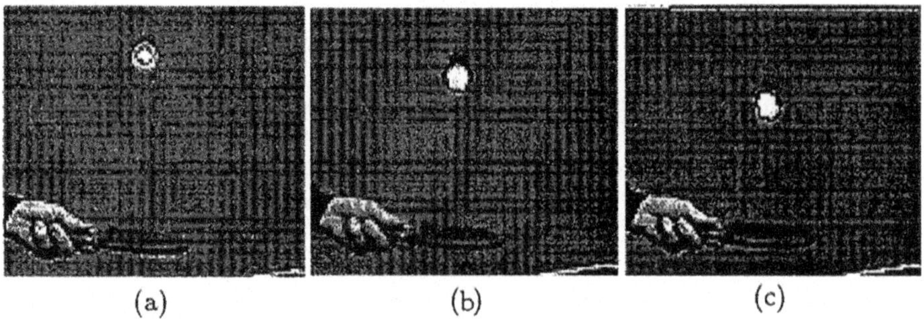

Fig. 6.1 Contour tracking example

$$E\left[C(p)\right] = \int_0^1 g(I_D(C(p)))\left|C'(p)\right|dp \tag{6.2}$$

where g represents a Gaussian function and I_D is the motion detection boundary image [20].

The moving regions of the image are successfully detected by this GAC model. But in order to track properly the video objects, the energy functional of the GAC is modified, by adding a video tracking term, as follows:

$$E\left[C(p)\right] = \int_0^1 \left(\underbrace{g(I_D(C(p)), \sigma_D)}_{Motion\ detection} + \underbrace{(1-\gamma)g(|\nabla I(C(p,t))|, \sigma_T)}_{Tracking\ term} \right) \left|C'(p)\right|dp \tag{6.3}$$

where the variances of the Gaussian functions, σ_D and σ_T, capture the motion detection and tracking information [20]. While the detection component forces the curve C to converge toward the moving region, the tracking term is used to evolve the curve until it reaches the right location of the moving object. A video object detection and tracking example based on this active contour-based framework is described in Fig. 6.2.

An object tracking technique based on morphing active contours was proposed by Bertalmio et al. in [21]. Their approach is based on deforming the object contours of interest from the first frame toward the desired location in the next one. A system of coupled nonlinear PDEs is used for this purpose [21]. It has the following form:

$$\begin{cases} \frac{\partial F_1(x,y,t)}{\partial t} = \beta(x,y,t)\|\nabla F_1(x,y,t)\| \\ \frac{\partial u(x,y,t)}{\partial t} = \widehat{\beta}(x,y,t)\|\nabla u(x,y,t)\| \end{cases} \tag{6.4}$$

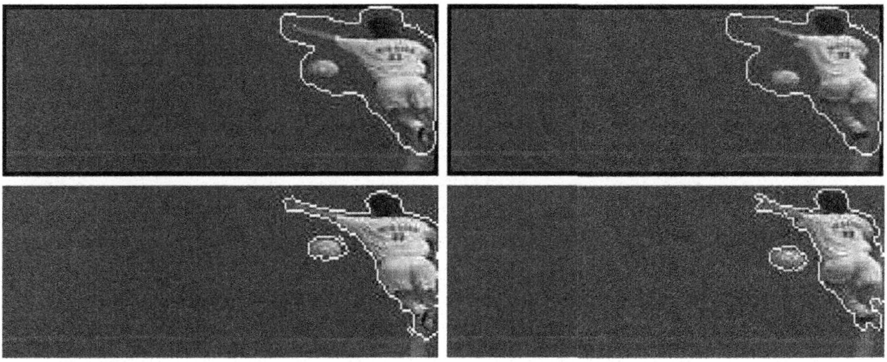

Fig. 6.2 Active contour-based video object detection and tracking example

where the images F_i represent features of the frames I_i (for example, the edge maps), u is a continuous and Lipschitz function whose zero level-set is the curve $C_{I_i}(p, 0)$, the velocity $\widehat{\beta}(x, y, t) = \beta(x, y, t) \cdot \vec{\mathcal{N}}_{F_1} \cdot \vec{\mathcal{N}}_u$, where β represents a function measuring the discrepancy between the features F_1 and F_2, and the normal

$$\begin{cases} \vec{\mathcal{N}}_{F_1} := \frac{\nabla F_1(x,y,t)}{\|\nabla F_1(x,y,t)\|} \\ \vec{\mathcal{N}}_u := \frac{\nabla u(x,y,t)}{\|\nabla u(x,y,t)\|} \end{cases} \tag{6.5}$$

The first equation in (6.4) performs the morphing process, which means deforms the first image, or its features, toward the second one. The second PDE represents the tracking equation. The object tracking is performed by projecting the velocities of the first PDE in (6.4) into the second one [21]. A video object tracking example illustrating this method is displayed in Fig. 6.3.

Another active contour-based multiple object tracking technique that tracks successfully non-rigid objects in color frames was introduced in [22]. It applies a two-step tracking process that is applied successively to each frame of the video sequence. The considered active contour represents a snake based on continuity, balloon, and curvature internal energies [22].

In the first step, the active contour is initialised in the current frame by using the result from the previous frame. The initialization is carried by a rectangle parallel to the image borders and surrounding the final active contour achieved at the previous frame [22].

Then, a contour deformation process is performed in the second step. Given the ballon force that is used here, the snake is forced to retract, instead of expand [22]. The position of the video object at time t is determined by using its previous location

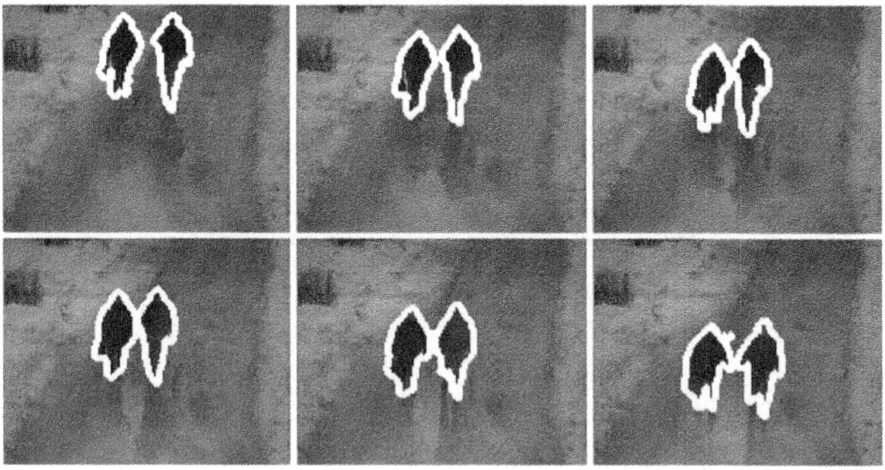

Fig. 6.3 Morphing active contour-based video object tracking

at time $t-1$. An active contour splitting process is also introduced in [22], in order to deal with the topology changes.

The geodesic active contour-based object detection algorithms developed by us, disseminated in [23, 24] and described in the Sect. 5.2.2 of the previous chapter could be also extended for the video tracking tasks. Thus, we consider modified versions of the PDE models given by (5.25) and (5.29), which are adapted for object tracking. For example, the GAC in (5.29) may lead to a tracking equation of the following type:

$$\frac{\partial u}{\partial t} - \alpha\varphi\left(\frac{\|\Delta u\| + \|\nabla u\|}{2}\right) div(\psi(t,x,y)\varphi(\|\nabla u\|)\nabla u) \\ - \beta\psi(t,x,y)\|\nabla u\| - (1-\psi(t,x,y)) \cdot \nabla u = 0 \tag{6.6}$$

6.2 Optical Flow-Based Moving Object Detection and Tracking Models

The video optical flow represents the pattern of apparent motion of various entities in a visual scene caused by the relative motion between the observer and that scene. It studies the relative motion of objects across different frame sequences using the velocity of the movement and illumination changes [25].

Optical flow represents an effective instrument for detecting the video objects and tracking their movement. Various optical flow estimation techniques have been developed in the last decades. These approaches could be grouped in several categories.

The block matching-based algorithms compute the optical flow by splitting the image into blocks and estimating the displacement to each block [26]. The parametric models cluster the pixels into regions and estimate the motion of these zones. The motion in these regions is parametrized, since one makes the assumption that the motion field in each region is fully characterised by a sequence of parameters. The purpose of a parametric model is to estimate the motion parameters that minimise a given loss function. A well-known parametric optical flow estimation model is the Lucas–Kanade algorithm that is based on rectangular regions and parameterises the motion as purely translational [27]. Deep learning-based techniques for optical flow computation have been also introduced. They include FlowNet [28] and other learning-based models based on CNNs [29]. Also, transformer-based optical flow estimation solutions have been introduced [30].

The regularized optical flow estimation models formulate optimization problems for this task. Thus, these PDE-based variational approaches formulate the flow as a global energy functional to be minimized:

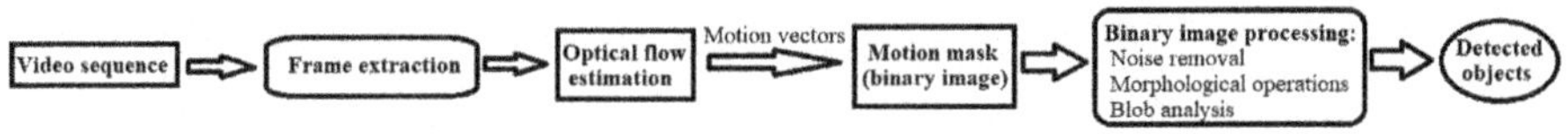

Fig. 6.4 Optical flow-based moving object detection scheme

$$E = \iint_{\Omega} \psi(I(x+u, y+v, t+1) - I(x, y, t)) + \alpha\psi(|\nabla u|) + \alpha\psi(|\nabla v|)dxdy \quad (6.7)$$

where ψ is a loss function that leads to the Horn–Schunck estimation model for $\psi(x) = x^2$ [31]. The functional of the Horn–Schunck model is minimized by solving the associated multi-dimensional Euler–Lagrange equations:

$$\begin{cases} I_x\left(I_x u + I_y v + I_t\right) - \alpha\Delta u = 0 \\ I_y\left(I_x u + I_y v + I_t\right) - \alpha\Delta v = 0 \end{cases} \quad (6.8)$$

Those partial differential equations are solved by applying the next iterative schemes, which determine the motion vectors [31]:

$$\begin{cases} u^{k+1} = \overline{u}^k - \frac{I_x\left(I_x\overline{u}^k + I_y\overline{v}^k + I_t\right)}{4\alpha^2 + I_x^2 + I_y^2} \\ v^{k+1} = \overline{v}^k - \frac{I_y\left(I_x\overline{u}^k + I_y\overline{v}^k + I_t\right)}{4\alpha^2 + I_x^2 + I_y^2} \end{cases} \quad (6.9)$$

The Horn–Schunck algorithm provides good motion estimation results, but it has a higher time complexity and it is also quite sensitive to noise. Other versions of the loss function, such as $\psi(x) = \sqrt{x^2 + \xi}$, could also be used in Eq. (6.7) [32].

The PDE-based optical flows can be applied successfully to the moving object detection tasks [32, 33]. Thus, the determined optical flow is used to detect the motion between two subsequent video frames. The general scheme of an optical flow-based video object detection process is described in Fig. 6.4.

First, the frames are extracted from the analyzed video. Then, an optical flow computation technique is applied to estimate the motion vectors in each video frame. A thresholding process that is performed on the determined motion vectors generates a binary image called *motion mask*. Several pre-processing operations are next applied to this binary image [33].

So, an image filtering process is performed to remove the scattered noise that is usually generated in the binary image by the background motion produced by slight camera movements and environmental conditions. The 2D median filtering represents an effective solution for these noises.

Then, several morphological operations are applied to the denoised binary image, using some properly chosen structuring elements [34]. Thus, some closing and eroding operations are performed on the motion mask to remove the small holes in blobs and the object regions which are unnecessarily detected as moving. A gap

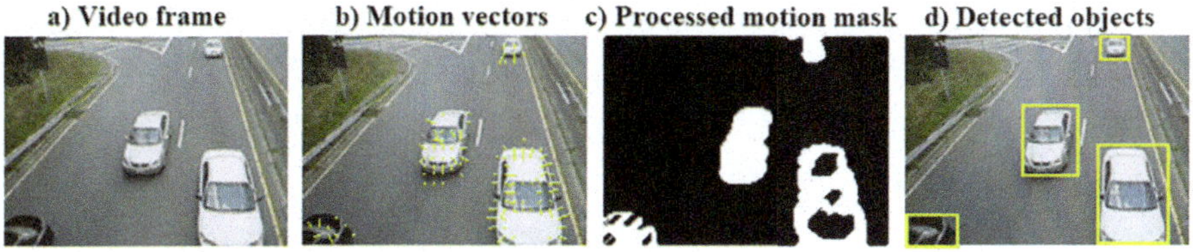

Fig. 6.5 Optical flow-based object detection process

filling operation could be also applied to the connected components of this binary image.

A blob analysis process that depends on the type of the objects considered for detection is then performed. Those connected components that do not correspond to actual objects have to be removed and only the blobs satisfying the moving objects characteristics must be preserved. Therefore, one should remove the blobs characterized by some inappropriate sizes, solidities and aspect (width to height) ratios. The remaining connected components represent the video frame foreground and correspond to the detected objects.

An example of a moving object detection process based on the optical flow estimation is described in Fig. 6.5. The objects subjected by this detection process are the moving cars in the video frame that is displayed in (a). Its motion vectors determined by the optical flow estimation approach are represented in (b). The motion mask processed by the mentioned operations is displayed in (c) and the detected vehicles, which correspond to its connected components, are depicted in the image (d).

Video object tracking task can also be successfully approached by using the optical flow estimation. The Horn–Schunck algorithm and its versions generate dense optical flow vectors that are applicable for tracking the objects moving with high speed [31, 35]. Since these video optical flow estimation methods can calculate properly the motion of each pixel between two successive frames, it can be used to determine the instance of each object in the next frame. Thus, the entire trajectory of an object in a video sequence could be identified by using the motion vectors generated by the optical flow.

Optical flow-based video object tracking represents an easy task for the static camera videos. The tracking task becomes more difficult for the objects in moving camera videos, since the camera motion is added to the local motion. The removal of the global motion from a video sequence can be performed by applying a normalized cross-correlation (NCC) algorithm that detects that motion [36]. Then, one subtracts the global motion for each image pixel, so that only the local motion remains. Next, the optical flow estimation can be performed properly on the sequence freed from the camera motion. A video object trajectory estimation based on global motion removal and a combination of optical flow and Kalman filtering is then provided in [36].

Another framework combines the optical flow to edge information to perform the object tracking [37]. The optical flow is determined, then the model detects a moving object by using it. Next, that object is tracked from the optical flow data, but in case

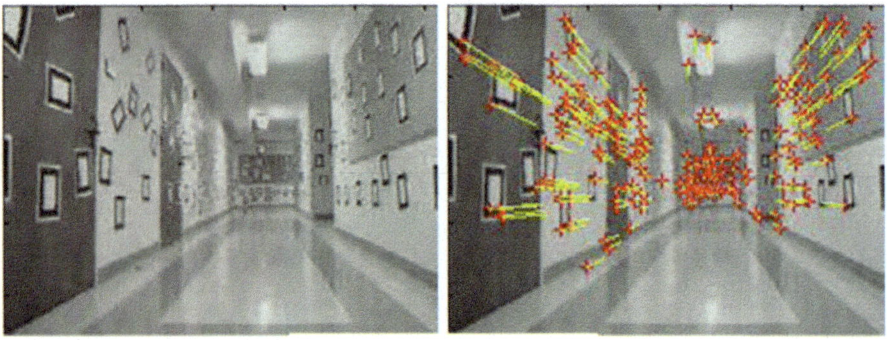

Fig. 6.6 Corner feature tracking example

of overlapping or stopping it cannot be tracked only with optical flows, so the edges are used to determine its contour [37].

The feature-based tracking techniques represent a category of effective methods based on optical flow estimation [38]. These approaches select a set of key points (pixels) from the subimage determined by the bounding box of a detected object and extract the visual features corresponding to them. These distinctive features are then tracked over multiple video frames by using the optical flow vectors determined by an estimation algorithm, thus enabling the identification of the object's instances across those frames.

Selecting the best interest keypoints and the visual features that could be tracked well still represents a quite difficult task. Some image features used for object tracking could represent Harris corners [39], textured areas, FAST features, windows with high spatial frequency content [40] and Shi–Tomasi corner features [38]. An example of a feature tracking process using corner features is described in Fig. 6.6.

We also considered optical flow-based tracking solutions for some classes of video objects [41]. Thus, the vehicle detection and tracking task was approached in [41] by using various machine and deep learning-based methods, including the optical flow estimation.

So, we applied an optical flow vector computation algorithm that was derived from the model (6.7) by introducing a new loss function ψ. Then, a feature tracking procedure using the obtained motion vectors was applied to the analyzed traffic video sequence [41]. The tracked features represented the keypoints determined by using the Shi–Tomashi corner detector [38]. Several frames from the video sequence representing that optical flow-based vehicle tracking process are described in Fig. 6.7.

Fig. 6.7 Optical flow-based vehicle tracking process

6.3 Video Object Tracking by Detection Using Anisotropic Diffusion-Based Multi-scale Image Analysis

The nonlinear PDE-based multi-scale analysis, which has been successfully applied to the image recognition, indexing and retrieval, and temporal video segmentation tasks described in previous two chapters, represents also an effective tool in the video object tracking domain. Thus, the powerful image content descriptors created by applying the multi-scale analysis could be used successfully by the tracking-by-detection techniques.

Tracking by detection (TBD) represents a class of video object tracking methods that first detects the objects of interest in each video frame and next associates those detections across frames. A TBD procedure matches the object detections by analyzing their appearance, location and/or motion features [42].

The appearance of an image object could be described properly by performing a high-level multi-scale feature extraction on it. We have developed several tracking-by-detection techniques applying nonlinear diffusion-based multi-scale feature extraction operations on the tracked objects and disseminated them in our research papers [24, 43, 44]. The video objects most commonly subjected to those tracking processes represent human persons and vehicles.

Thus, a multiple vehicle detection and tracking approach is provided by us in [43]. A combined deep and machine learning vehicle localization technique is applied in the detection stage. First, it performs a GMM-based background subtraction on the analyzed frame [10]. Next, the obtained GMM-based moving regions are combined to the detection results generated by some deep networks trained on vehicle datasets, such as Faster R-CNN and YOLO-v2 [11], and a trained Aggregated Channels Network (ACF)-based detector [9].

Then, a tracking by detection approach using a nonlinear reaction–diffusion based multi-scale analysis applied to the detected vehicles is proposed in [43]. A scale-space representation is built there using the following nonlinear fourth-order anisotropic diffusion-based filtering model:

$$\begin{cases} \frac{\partial u}{\partial t} + \lambda\xi\left(\left\|\nabla^2 u\right\|\right)\Delta\left(\psi(\|\nabla u_\sigma\|)\nabla^2 u\right) + \alpha(u - u_0) = 0, & \text{in } (0,T) \times \Omega \\ u(0, x, y) = u_0(x, y), \quad \forall(x,y) \in \Omega \\ u(t, x, y) = 0, \quad \forall(x,y) \in \partial\Omega, & \text{in}(0,T) \times \Omega \\ \frac{\partial}{\partial \vec{n}}(t, x, y) = 0, \quad \forall(x,y) \in \partial\Omega, & \text{in}(0,T) \times \Omega \end{cases} \tag{6.10}$$

with $\lambda \in [1, 2)$, $\alpha \in (0, 0.5)$, $\Omega \subseteq R^2$, $u_0 \in L^2(\Omega)$ and $u_\sigma = u * G_\sigma$, where the 2D Gaussian filter kernel $G_\sigma(x, y) = \frac{1}{2\pi\sigma^2}e^{-\frac{x^2+y^2}{2\sigma^2}}$. The following positive and monotonic decreasing diffusivity function is considered [43]:

$$\psi : [0, \infty) \to [0, \infty) : \psi(s) = \beta\sqrt[3]{\frac{\delta}{\left|\zeta + \gamma s^2\right|}} \tag{6.11}$$

where $\gamma \in (0, 1]$, $\zeta, \delta \geq 5$, $\beta \geq 1$. The term controlling the speed of this diffusion process is based on the next function:

$$\xi : [0, \infty) \to [0, \infty) : \xi(s) = \varsigma(\eta s^\varepsilon + \nu)^{\frac{1}{\varepsilon+1}} \tag{6.12}$$

where $\nu, \eta \in [1, 5)$, $\varsigma \in (0, 1)$ and $\varepsilon \in (0, 1)$ [43].

The nonlinear fourth-order PDE model (6.10) is non-variational, since it cannot be derived applying an energy cost functional minimization and it is also well-posed. A rigorous mathematical investigation is performed on it in [43], to demonstrate its validity (well-posedness). We demonstrate it admits at least one variational (weak) solution that is a function $u : (0, T) \times \Omega \to \mathbb{R}$ satisfying the next conditions:

$$u \in C\left([0, T]; L^2(\Omega)\right) \cap L^2\left(0, T; H^2(\Omega)\right), \quad \frac{du}{dt} \in L^2(0, T; V') \tag{6.13}$$

and

$$\begin{aligned} &\frac{d}{dt}\int_\Omega u(t, x, y)\varphi(x, y)dxdy \\ &+ \lambda\int_\Omega \psi(|\nabla u(t, x, y)|)\Delta u(t, x, y)\cdot\Delta(\xi(|\nabla u(t, x, y)|)\varphi(x, y))dxdy \\ &+ \alpha\int_\Omega (u(t, x, y) - u_0(x, y))\varphi(x, y)dxdy = 0, \quad \text{a.e. } t \in (0, T) \end{aligned} \tag{6.14}$$

for $\forall\varphi \in V = \left\{z \in H^2(\Omega) \cap H_0^1(\Omega); \frac{\partial z}{\partial \vec{n}} = 0 \text{ on } \partial\Omega\right\}$. We have $V \subset H^2(\Omega) \subset V'$ (dual of V).

In [43] we denote for $\forall v \in L^2(0, T; V)$ the operator $A^v : V \to V'$, where

$$_{V'}\left(A^{v}(u),\varphi\right)_{V}=\lambda\int\limits_{\Omega}\psi(|\nabla v|)\Delta u\Delta(\xi(|\nabla v|)v)dxdy+\alpha\int\limits_{\Omega}(u-u_0)\varphi dxdy \quad (6.15)$$

and set $\kappa=\left\{v\in L^2(0,T;V);\ \int_0^T\|v(t,x,y)\|_V^2\le q^2\right\}$, $q>0$—a constant. For $v\in\kappa$, A^v is continuous from V to V' and $_{V'}(A^v(u),u)_V\ge c\|u\|_V^2$, $\forall u\in V$, $c>0$. So, by the standard existence theory [45], for $\forall v\in\kappa$ there is a unique solution $u=\phi(v)\in\kappa$ to the equation:

$$\begin{cases}\frac{du}{dt}+A^v(u)=0, & t\in(0,T)\\ u(0)=u_0\end{cases} \quad (6.16)$$

Further more, $v\to\phi(v)$ maps κ into itself, is continuous and compact [45]. Next, by applying the Schauder's fixed-point theorem it follows that there is $u\in\kappa$ such that $\phi(u)=u$, and that u represents a weak solution to (6.10). In general its uniqueness does not result under the present conditions, but a local uniqueness result could be achieved for T sufficiently small.

That solution is computed numerically by using a consistent and stable approximation scheme that is proposed in [43] and converges fast to it. The next explicit iterative numerical algorithm is obtained by applying the finite differences on (6.10) [46]:

$$u_{i,j}^{n+1}=u_{i,j}^{n}(\alpha+1)-\lambda\xi\left(\left\|\Delta u_{i,j}^{n}\right\|\right)\left(\psi_{i+1,j}^{n}+\psi_{i-1,j}^{n}+\psi_{i,j+1}^{n}+\psi_{i,j-1}^{n}-4\psi_{i,j}^{n}\right)-u_{i,j}^{0}\alpha \quad (6.17)$$

where $\psi_{i,j}^{n}=\psi\left(\left|\nabla u_{i,j}^{n}\right|\right)\Delta u_{i,j}^{n}$, $\Delta u_{i,j}^{n}=u_{i+1,j}^{n}+u_{i-1,j}^{n}+u_{i,j+1}^{n}+u_{i,j-1}^{n}-4u_{i,j}$ and $x=ih$, $y=jh$, $i\in\{1,\ldots,I\}$, $j\in\{1,\ldots,J\}$ for a $[Ih\times Jh]$ support image.

The numerical approximation scheme (6.17) is then used to create a multi-scale space for the vehicle feature extraction process. The scale-space representation is created in the same way as in the Sects. 4.3 and 5.3, treating the multi-scale based image indexing and retrieval and temporal video segmentation. So, the following multi-scale representation is obtained for the vehicle object:

$$\begin{aligned}S(Veh)=\big\{&Veh, RGB\left(\left[(L(Veh))^{\tau},a(Veh),b(Veh)\right]\right),\ldots,\\ &RGB\left(\left[(L(Veh))^{\tau(K-1)},a(Veh),b(Veh)\right]\right)\big\}\end{aligned} \quad (6.18)$$

where $K\ge 3$, $\left[(L(Veh))^{\tau},a(Veh),b(Veh)\right]$ represents the vehicle sub-image converted in L*a*b* and with the luminance channel filtered with τ iterations of (6.17), with $\tau\in[4,10]$ [43].

A content-based feature extraction using scale-space S is performed on all detected vehicles. A high-level multi-scale analysis is applied to each vehicle Veh_j^i, where $i\in\{1,\ldots,M\}$, $j\in\{1,\ldots,n(F_i)\}$, where M is the number of frames and $n(F_i)$ is the number of detected vehicles in the ith frame. A high-level feature extraction based on SURF and HOG-based characteristics is applied on the vehicle image at each scale

$k \in \{0, \ldots, K-1\}$, $S\left(Veh_j^i\right)\{k\} = RGB\left(\left[\left(L\left(Veh_j^i\right)\right)^{\tau k}, a\left(Veh_j^i\right), b\left(Veh_j^i\right)\right]\right)$. A 2D feature vector is obtained for it by applying the next matrix concatenation:

$$V\left(S\left(Veh_j^i\right)\{k\}\right) := \left[V_{SURF}\left(S\left(Veh_j^i\right)\{k\}\right) \; V_{HOG}\left(S\left(Veh_j^i\right)\{k\}\right)\right] \tag{6.19}$$

where $V_{SURF}\left(S\left(Veh_j^i\right)\{k\}\right)$ represents the $[N_{SURF} \times 64]$ matrix having the feature vectors of 64 coefficients of the strongest $N_{\text{SURF}} > 5$ SURF keypoints as rows [7], and $V_{HOG}\left(S\left(Veh_j^i\right)\{k\}\right)$ is the $[N_{SURF} \times 36]$ matrix containing the HOG feature vectors of 36 coefficients [8] computed around the locations of those N_{SURF} keypoints [43].

Then, all the $[N_{SURF} \times 100]$ 2D feature vectors determined at multiple scales of S are concatenated into the final vehicle feature vector, as follows:

$$V\left(Veh_j^i\right) := \left[V\left(S\left(Veh_j^i\right)\{0\}\right) \; V\left(S\left(Veh_j^i\right)\{1\}\right) \; \ldots \; V\left(S\left(Veh_j^i\right)\{K-1\}\right)\right] \tag{6.20}$$

The $[N_{SURF} \times 100K]$ feature vectors $V\left(Veh_j^i\right)$ represent powerful content descriptors of the vehicle detections that are successfully used in the tracking process. Our vehicle tracking by detection approach matches the vehicle instances on successive frames by using the distances between these feature vectors in combination with the distances between vehicle centroids and some size-related measures [43]. Thus, the vehicle matching process between the frames F_i and F_{i+1} is formalized as:

$$match_i\left(\arg \min_{j \in UM^i(F_i), k \in UM^i(F_{i+1})} \left\{ d_{jk}^i \leq t_{fv} \,\middle|\, d\left(Veh_j^i, Veh_k^{i+1}\right) \leq t_{veh}, \frac{\max\left(Area\left(Veh_j^i\right), Area\left(Veh_k^{i+1}\right)\right)}{\min\left(Area\left(Veh_j^i\right), Area\left(Veh_k^{i+1}\right)\right)} \leq t_{size} \right\}\right) := 1 \tag{6.21}$$

where $match_i(j,k) = 1$ if Veh_j^i and Veh_k^{i+1} are matched, the set $UM^i(F_i) \subseteq \{1, \ldots, n(F_i)\}$ contains the indices of still unmatched vehicles in F_i, $d\left(Veh_j^i, Veh_k^{i+1}\right)$ is the distance between the centers of those vehicles, $Area\left(Veh_j^i\right)$ is the vehicle image area and $d_{jk}^i = d\left(V\left(Veh_j^i\right), V\left(Veh_k^{i+1}\right)\right)$, $\forall i \in \{1, \ldots, M\}, j \in \{1, \ldots, n(F_i)\}, k \in \{1, \ldots, n(F_{i+1})\}$.

The vehicle tracking (counting) algorithm determines all the matchings between any two successive frames of the video, by applying (6.21). Then, the tracking results, representing the vehicle trajectories, are achieved by combining all matchings [43].

A vehicle tracking example based on this TBD technique is described in Fig. 6.8. The trajectories of 2 detected cars are indicated by the red and yellow colors of their bounding boxes and feature vector distances d_{jk}^i are attached to the corresponding boxes in each F_i. The first (red) value of each vehicle is the feature vector distance to the first vehicle (red box) in the next frame, while the second (black) one is the feature vector distance to the second vehicle (yellow box). The lower value leads to the vehicle's next instance [43].

This multiple vehicle detection and tracking framework was tested successfully on many video datasets, the most important of them being UA-DETRAC database [47]. It works properly in both clean and noisy conditions and it has a quite high complexity, given the high computational costs of the machine and deep learning based object detection and the multi-scale feature extraction. However, its running time depends on the video's length M, its frames' sizes and the number of the vehicles in these frames.

Its performance has been evaluated on the UA-DETRAC multi-object detection and tracking benchmark containing 10 h of videos recorded at 25 fps, with resolution of [960 × 540] pixels. That dataset has more than 140K captured video frames and 1.21M labeled vehicle bounding boxes. The detection component of this framework outperforms each of the detection approaches it combines (GMM, ACF, Faster R-CNN and YOLO-v2) taken separately, and other machine and deep learning-based detectors also, as shown by the method comparison results discussed in [43].

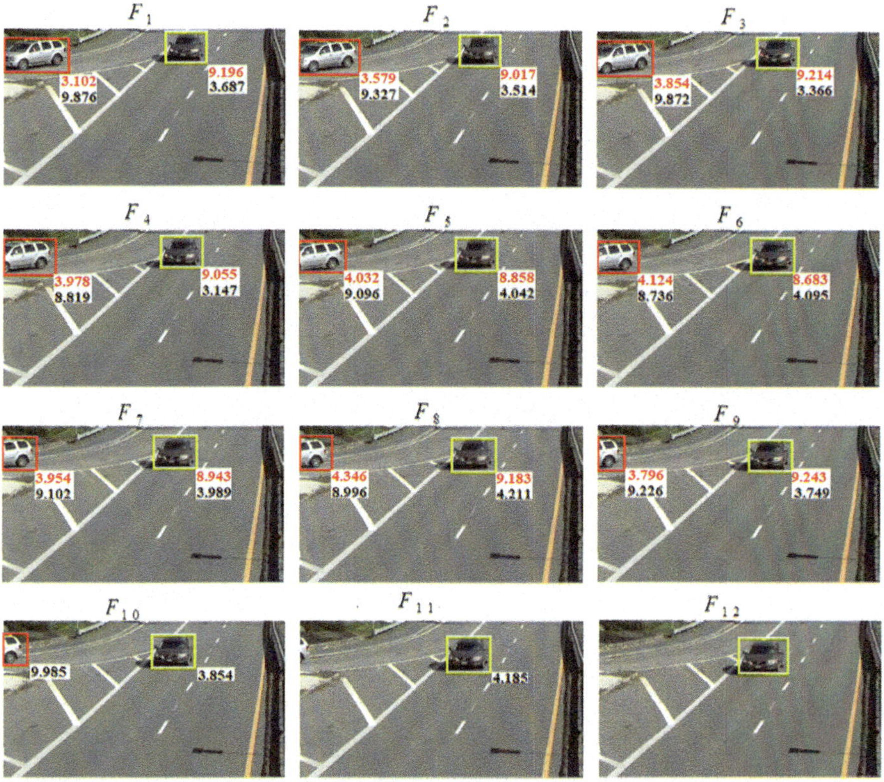

Fig. 6.8 A TBD-based vehicle counting process

Table 6.1 Multiple vehicle tracking method comparison

Tracking technique	Precision	Recall
The proposed technique	0.8401	0.8216
GMM + mean-shift tracking	0.7148	0.7332
GMM + Kalman filter	0.7643	0.7527
EB + IoU tracker	0.8576	0.8615
SIFT tracking	0.6233	0.6181
SIFT + Kalman filtering	0.7486	0.7391
Frame differencing + object matching	0.5801	0.5879
HOG + SVM + feature matching	0.6102	0.6213

Our nonlinear PDE-based vehicle tracking technique gets high values of the performance metrics and outperforms other counting methods, as illustrated by the vehicle tracking method comparison results described in Table 6.1. It is slightly outperformed only by the deep learning framework using the Evolving Boxes (EB)-based vehicle detection with IoU tracker [48].

Another video object detection and tracking technique using a nonlinear anisotropic diffusion-based multi-scale image analysis is proposed by us in [44]. Unlike the previous approach, it deals with moving people only and applies a combination of machine learhing schemes in the detection stage and a deep learning-based feature extraction in the tracking stage.

So, this pedestrian detection approach combines the results achieved by a person detector based on HOG and SVM [8], to those obtained by an ACF trained on the voluminous Caltech Pedestrian dataset [49]. Then a GMM-based background subtraction is applied on the video frame to detect the moving regions [44]. The obtained GMM-based frame foreground is combined to the SVM + HOG-based and ACF-based people detection results by applying a combining scheme described in [44]. A set of pedestrian detections, $P_i = \{p_1^i, \ldots, p_{n_i}^i\}$, is determined for each frame F_i and a tracking by detection algorithm is then applied on them.

The following second-order nonlinear anisotropic diffusion model is proposed there for the multi-scale pedestrian object analysis:

$$\begin{cases} \frac{\partial u}{\partial t} - \alpha\psi\left(\left|\nabla^2 u\right|\right)\nabla \cdot (\delta(\|\nabla u\|)\nabla u) + \beta(u - u_0) = 0, \quad (x,y) \in \Omega \\ u(0, x, y) = u_0(x, y), \quad \forall(x,y) \in \Omega \\ u(t, x, y) = 0, \quad \forall(x,y) \in \partial\Omega; \end{cases} \tag{6.22}$$

where $\alpha \in [1, 2), \beta \in (0, 0.4), \Omega \subseteq R^2$ and the observation $u_0 \in L^2(\Omega)$. One considers the next positive and monotonic decreasing edge-stopping function:

$$\delta : [0, \infty) \to [0, \infty) : \delta(s) = \lambda\left(\frac{\xi}{\left|\gamma s^3 + \eta \log_{10} \xi\right|}\right)^{\frac{1}{2}} \tag{6.23}$$

where $\eta \in (0, 1]$, $\gamma \geq 3$, $\xi \geq 4$ and $\lambda \geq 1$. The other positive function has the form:

$$\psi : [0, \infty) \to [0, \infty) : \psi(s) = \zeta \sqrt[r]{\varphi s^r + \nu}, \tag{6.24}$$

where $\varphi, \nu \in [1, 5)$, $\zeta \in (0, 0.5, 1)$ and $r \in (0, 1)$.

This second-order parabolic PDE-based model is non-variational and well-posed [44]. Its solution is determined numerically applying a finite difference-based approximation algorithm to (6.24) [46]. The next explicit iterative numerical approximation scheme that is consistent to that PDE is obtained in [44]:

$$\begin{aligned} u_{i,j}^{n+1} &= u_{i,j}^n(1-\beta) + u_{i,j}^0\beta + \alpha\psi\left(u_{i+h,j}^n + u_{i-h,j}^n + u_{i,j+h}^n + u_{i,j-h}^n - 4u_{i,j}\right) \\ &\cdot \left(\delta_{i+\frac{1}{2},j}\left(u_{i+1,j}^n - u_{i,j}^n\right) - \delta_{i-\frac{h}{2},j}\left(u_{i,j}^n - u_{i-1,j}^n\right)\right. \\ &\left.+\delta_{i,j+\frac{1}{2}}\left(u_{i,j+1}^n - u_{i,j}^n\right) - \delta_{i,j-\frac{1}{2}}\left(u_{i,j}^n - u_{i,j-1}^n\right)\right) \end{aligned} \tag{6.25}$$

where $\delta_{i\pm\frac{h}{2},j} = \frac{\delta_{i\pm h,j}+\delta_{i,j}}{2}$, $\delta_{i,j\pm\frac{h}{2}} = \frac{\delta_{i,j\pm h}+\delta_{i,j}}{2}$ [44].

The next scale-space representation is created, in the same way as in the previous case, for each pedestrian $p \in P_i$ by applying (6.25):

$$S(p) = \left\{p, RGB\left(\left[(L(p))^{\rho}, a(p), b(p)\right]\right), \ldots, RGB\left(\left[(L(p))^{\rho(K-1)}, a(p), b(p)\right]\right)\right\} \tag{6.26}$$

where $K \geq 3$, iteration step $\rho \in [3, 10]$ and $RGB(\ )$ provides the RGB reconversion. A high-level deep learning-based multi-scale feature extraction is performed on each pedestrian image p_j^i, with $i \in \{1, \ldots, M\}$ and $j \in \{1, \ldots, n_i\}$, using the scale-space S [44]. Two convolutional neural networks are used to extract the high-level content features of its image at kth scale, $S\left(p_j^i\right)\{k\} = RGB\left(\left[\left(L\left(p_j^i\right)\right)^{\rho k}, a\left(p_j^i\right), b\left(p_j^i\right)\right]\right)$.

A 2D feature vector is determined for $S\left(p_j^i\right)\{k\}$ as the concatenation $V\left(S\left(p_j^i\right)\{k\}\right) := \left[V_1\left(S\left(p_j^i\right)\{k\}\right); V_2\left(S\left(p_j^i\right)\{k\}\right)\right]$, where $V_1\left(S\left(p_j^i\right)\{k\}\right)$ is the feature vector with 1000 coefficients obtained by activating the fully connected *prediction* layer of a Inception-V3 network on the image version that is normalized and pre-processed according to Inception input layer specifications [50], and $V_2\left(S\left(p_j^i\right)\{k\}\right)$ represents the feature vector with 1000 coefficients achieved by activating the FC 1000 layer of a ResNet-101 model on the image version that is normalized and pre-processed according to the residual network' input layer specifications [51]. All the 2D feature vectors of the scale space are concatenated into the final pedestrian image feature vector: $V\left(p_j^i\right) := \left[V\left(S\left(p_j^i\right)\{0\}\right) \ldots V\left(S\left(p_j^i\right)\{K-1\}\right)\right]$. The $[2 \times 1000K]$ feature vectors $V\left(p_j^i\right)$ represent effective content descriptor of the pedestrian objects p_j^i that are used successfully in the people tracking by detection process [44].

The proposed pedestrian tracking algorithm determines the correspondences between the pedestrian detection located on consecutive frames by computing the minimum distances between their feature vectors. Thus, for each pedestrian p_j^i in F_i, one searches in a neighbourhood of it in F_{i+1} for those unmatched pedestrians that do not differ much in sizes and selects that corresponding to the minimum distance, which is p_κ^{i+1} where $\kappa = \arg\min_{k \in \{1,\dots,n_{i+1}\}} d\left(V\left(p_j^i\right), V\left(p_k^{i+1}\right)\right)$, as its next instance. Our TBD algorithm described in detail in [44] determines the tracks of all pedestrians in the analyzed video by combining all these matchings between the detections.

A pedestrian tracking example based on this approach is provided in Fig. 6.9. A set of values representing the feature vector distances to the detections in next frame is associated to each pedestrian detection and the minimum values correspond to its next instance. The determined tracks are marked by red and blue colors [44].

This multi-pedestrian detection and tracking approach was been tested successfully on many video sequences with tens of thousands of frames containing front-view, rear-view, left-view and right-view pedestrians, from several pedestrian video databases, such as the pedestrian movement direction recognition dataset of Dominguez-Sanchez et al. [52] and JAAD Dataset [53].

Our framework obtains a high detection and tracking accuracy. It generates a low number of false positives and negatives, which leads to high *Precision* and *Recall* scores. These performance metric scores are presented in Table 6.2 describing the method comparison results. This tracking-by-detection technique achieves high values of these measures, outperforming other machine and deep learning approaches [44].

The deep learning-based multi-scale feature extraction is also used in the tracking stage of the framework disseminated by us in [24], whose detection component, based on a GAD Haar Cascade Classifier trained in OpenCV and a GAC-based segmentation, was already discussed in Sect. 5.2.2.

The scale-space representation is created by applying the following second-order reaction–diffusion model for the image function of the analyzed vehicle v:

$$\begin{cases} \frac{\partial v}{\partial t} - \alpha \nabla \cdot (\varphi(\|\nabla v\|)\nabla v_\sigma)\delta\left(|\nabla v| + \left|\nabla^2 v\right|\right) + \varepsilon(v - v_0) = 0 \\ v(x, y, 0) = v_0(x, y), \quad \forall (x,y) \in \Omega \\ v(x, y, t) = 0, \quad \forall (x,y) \in \partial\Omega, \quad t \in (0,T] \end{cases} \tag{6.27}$$

where $\varepsilon \in (0, 1)$, its diffusivity function $\varphi(s) = \zeta\left(\frac{\lambda}{|\xi s^\kappa + \eta|}\right)^{\frac{1}{\kappa+1}}$ and $\delta(s) = \zeta\sqrt[3]{\rho s^2 + r}$, with $r, \rho \in (1, 5]$ [24].

This nonlinear PDE-based model is solved numerically by applying a finite difference-based approximation algorithm developed by us [46]. First, the component $\frac{\partial v}{\partial t} + \varepsilon(v - v_0)$ is discretized as:

$$\frac{v_{i,j}^{n+\Delta t} - v_{i,j}^n}{\Delta t} + \varepsilon\left(v_{i,j}^n - v_{i,j}^0\right) = v_{i,j}^{n+\Delta t}\frac{1}{\Delta t} + v_{i,j}^n\left(\varepsilon - \frac{1}{\Delta t}\right) - \varepsilon v_{i,j}^0, \tag{6.28}$$

Fig. 6.9 Pedestrian tracking process

Table 6.2 Multi-pedestrian tracking method comparison

Technique	Precision	Recall
The proposed framework	0.8467	0.8351
AdaBoost + block matching	0.8312	0.8221
Faster R-CNN + color feature matching	0.7839	0.7693
HOG + SVM + Kalman filtering	0.7362	0.7285
Temporal differencing + object matching	0.7137	0.7214
HOG + SVM + feature matching	0.6423	0.6372

then $\delta\left(|\nabla v| + \left|\nabla^2 v\right|\right)$ is discretized as $\delta\left(\sqrt{\left(\frac{v^n_{i+h,j}-v^n_{i-h,j}}{2h}\right)^2 + \left(\frac{v^n_{i,j+h}-v^n_{i,j-h}}{2h}\right)^2} + \frac{v^n_{i+h,j}+v^n_{i-h,j}+v^n_{i,j+h}+v^n_{i,j-h}-4v^n_{i,j}}{h^2}\right)$ and the divergence component $\nabla \cdot (\varphi(\|\nabla v\|)\nabla v_\sigma)$ is approximated numerically as

$$\frac{\varphi\left(\left\|v^n_{i+h,j}-v^n_{i,j}\right\|\right)\left((v_\sigma)^n_{i+h,j}-(v_\sigma)^n_{i,j}\right)+\varphi\left(\left\|v^n_{i-h,j}-v^n_{i,j}\right\|\right)\left((v_\sigma)^n_{i-h,j}-(v_\sigma)^n_{i,j}\right)+\varphi\left(\left\|v^n_{i,j+h}-v^n_{i,j}\right\|\right)\left((v_\sigma)^n_{i,j+h}-(v_\sigma)^n_{i,j}\right)+\varphi\left(\left\|v^n_{i,j-h}-v^n_{i,j}\right\|\right)\left((v_\sigma)^n_{i,j-h}-(v_\sigma)^n_{i,j}\right)}{4},$$

where $t = n\Delta t, n \in \{0, \ldots, N\}$ and $x = ih, y = jh, i \in \{1, \ldots, I\}, j \in \{1, \ldots, J\}$ [24]. One considers $h = \Delta t = 1$ and get the next explicit iterative numerical approximation scheme:

$$v^{n+1}_{i,j} = v^n_{i,j}(1-\varepsilon) + \delta\left(\frac{\sqrt{\left(v^n_{i+1,j}-v^n_{i-1,j}\right)^2 + \left(v^n_{i,j+1}-v^n_{i,j-1}\right)^2}}{2} + v^n_{i+1,j} + v^n_{i-1,j} + v^n_{i,j+1} + v^n_{i,j-1} - 4v^n_{i,j}\right)$$
$$\cdot \frac{\alpha}{4}\begin{bmatrix} \varphi\left(\left\|v^n_{i+1,j}-v^n_{i,j}\right\|\right)\left((v_\sigma)^n_{i+1,j}-(v_\sigma)^n_{i,j}\right)+\varphi\left(\left\|v^n_{i-1,j}-v^n_{i,j}\right\|\right)\left((v_\sigma)^n_{i-1,j}-(v_\sigma)^n_{i,j}\right)+ \\ \varphi\left(\left\|v^n_{i,j+1}-v^n_{i,j}\right\|\right)\left((v_\sigma)^n_{i,j+1}-(v_\sigma)^n_{i,j}\right)+\varphi\left(\left\|v^n_{i,j-1}-v^n_{i,j}\right\|\right)\left((v_\sigma)^n_{i,j-1}-(v_\sigma)^n_{i,j}\right)\end{bmatrix} + \varepsilon v^0_{i,j} \tag{6.29}$$

This fast-converging numerical algorithm that is consistent to the reaction–diffusion model (6.27), is then used to create the multi-scale space representation of the vehicle detection v. Therefore, one obtains the scale-space with K scale $\left\{v^0, v^5, \ldots, v^{5(K-1)}\right\}$, where each v^{5k} is computed by (6.29) [24].

A deep learning-based feature extraction is performed on the vehicle at each scale. The approach extracts its high-level characteristics by activating the deep layers of 2 CNNs. One obtains the feature vector $V\left(v^{5k}\right) := \left[V_1\left(v^{5k}\right); V_2\left(v^{5k}\right)\right]$, where $V_1\left(v^{5k}\right)$ is generated by an activation of the FC 1000 layer of ResNet-50 on v^{5k} and $V_2\left(v^{5k}\right)$ is produced by an activation of the fully-connected layer, *Logits*, of a MobileNet-V2 model assembled by us [24]. The final high-level vehicle feature vector is achieved by concatenating these [2 × 1000] feature vectors achieved at multiple scales, as follows: $V(v) := \left[V\left(v^0\right), V\left(v^5\right), \ldots, V\left(v^{5k}\right), \ldots, V\left(v^{5(K-1)}\right)\right]$ [24].

These deep learning-based vehicle feature vectors are applied effectively in by the TBD algorithm of this framework, which is detailed in [24]. The next instance of v_j^i is the detection v_τ^{i+1} from the next frame satisfying some location and size conditions and corresponding to the minimum feature vector distance, therefore $\tau = \arg\min_k d_E\left(V\left(v_j^i\right), V\left(v_k^{i+1}\right)\right)$, where d_E is the Euclidean distance [24].

A vehicle tracking by detection example using this technique is described in Fig. 6.10. The cascade classifier-based vehicle detections are displayed in (a). The detections improved by our GAC-based segmentation approach are described in (b). The vehicle instance matching process based on the DL-based feature vector distance values is presented in (c) and (d) [24].

We performed many experiments and method comparisons based on this framework. Being motion-insensitive, it detects properly both the static and moving vehicles and works successfully for both stationary and moving-camera videos. It generates only few false positives and negatives for the detection and tracking tasks, getting

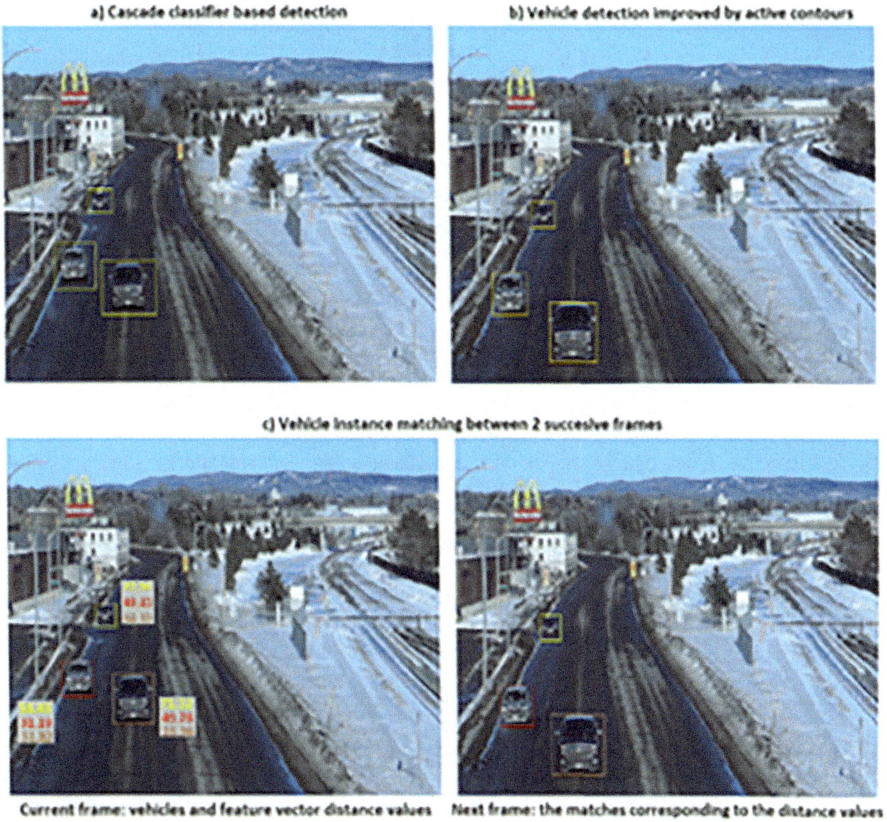

Fig. 6.10 Vehicle tracking by-detection example

Table 6.3 Multi-vehicle detection and tracking method comparison

Technique	Precision	Recall
The proposed approach	0.8541	0.8432
Temporal differencing + object matching	0.7525	0.7416
ACF + IoU	0.8223	0.8137
Faster R-CNN + IoU tracker	0.8429	0.8391
GMM + Kalman filtering	0.7362	0.7285
YOLO-Net + DeepSORT	0.9108	0.9012

high scores *Precision* and *Recall*, as illustrated by the results described in Table 6.3 [24].

As shown by these scores, the described technique outperforms some machine and deep learning-based vehicle detection and tracking algorithms [2, 9–11], but it is outperformed by a framework combining the YOLO detection to DeepSORT tracking [54].

References

1. A. Banharnsakun, S. Tanathong, Object detection based on template matching through use of best-so-far ABC, *Computational intelligence and neuroscience*, 2014.
2. T. Barbu, Multiple Object Detection and Tracking in Sonar Movies using an Improved Temporal Differencing Approach and Texture Analysis, *U.P.B. Scientific Bulletin*, Series A, 74, pp. 27–40, 2012.
3. A. Wu, S. Zhao, C. Deng, W. Liu, Generalized and Discriminative Few-Shot Object Detection via SVD-Dictionary Enhancement, *Advances in Neural Information Processing Systems*, 34, 2021.
4. T. Mordan, N. Thome, G. Henaff, M. Cord, End-to-end learning of latent deformable part-based representations for object detection, *International Journal of Computer Vision*, 127, pp. 1659–1679, 2019.
5. P. Viola, M. Jones, Rapid object detection using a boosted cascade of simple features, *Proceedings of CVPR 2001, 2001. IEEE.*
6. T. Nguyen, E. Park, J. Han, D. C. Park, S. Y. Min, Object detection using scale invariant feature transform, *Genetic and evolutionary computing*, pp. 65–72, Springer, Cham, 2014.
7. J. Farooq, Object detection and identification using SURF and BoW model, *Intl. Conf. on Computing, Electronic and Electrical Engineering (ICE Cube)*, pp. 318–323, April 2016.
8. N. Dalal, B. Triggs, Histograms of oriented gradients for human detection, *IEEE computer society conference on computer vision and pattern recognition (CVPR'05)*, Vol. 1, pp. 886–893, June 2005.
9. J. Hua, Y. Shi, C. Xie, H. Zhang and J. Zhang, Pedestrian-and vehicle-detection algorithm based on improved aggregated channel features, *IEEE Access*, 9, pp. 25885–25897, 2021.
10. R. Y. Bakti, I. S. Areni, A. Prayogi, Vehicle detection and tracking using gaussian mixture model and kalman filter, *International Conf on Computational Intelligence and Cybernetics*, pp. 115–119, 2016. IEEE.
11. L. Liu, et al., *Deep learning for generic object detection: A survey*, International journal of computer vision, 128 (2), pp. 261–318, 2020.
12. X. Li, K. Wang, A multiple object tracking method using Kalman filter, *IEEE international conference on information and automation*, July 2010.

13. C. Yang, R. Duraiswami, L. Davis, Efficient mean-shift tracking via a new similarity measure, *IEEE Conference on Computer Vision and Pattern Recognition (CVPR'05)*, Vol. 1, pp. 176–183, 2005. IEEE.
14. Y. Yuan, H., Yang, Y. Fang, W. Lin, Visual object tracking by structure complexity coefficients, *IEEE Transactions on Multimedia*, 17(8), pp. 1125–1136, 2015.
15. V. Srikrishnan, T. Nagaraj, S. Chaudhuri, Fragment based tracking for scale and orientation adaptation", *6th Indian Conf. on Computer Vision, Graphics & Image Processing*, pp. 328–335, Dec. 2008. IEEE.
16. C. C. Hsu, G. T. Dai, Multiple object tracking using particle swarm optimization, *International Journal of Electronics and Communication Engineering*, 6(8), pp. 744–747, 2012.
17. G. Ciaparrone, et al., Deep learning in video multi-object tracking: A survey, *Neurocomputing*, 381, pp. 61–88, 2020.
18. V. Caselles, R. Kimmel, G. Sapiro, Geodesic active contours, *International journal of computer vision*, 22, pp. 61–79, 1997.
19. V. Caselles, B. Coll, Snakes in movement, *SIAM Journal on Numerical Analysis*, *33*(6), pp. 2445–2456, 1996.
20. N. Paragios, R. Deriche, Geodesic active contours and level sets for the detection and tracking of moving objects, *IEEE Transactions on pattern analysis and machine intelligence*, *22*(3), pp. 266–280, 2000.
21. M. Bertalmio, G. Sapiro, G. Randall, Morphing active contours, *IEEE Transactions on Pattern Analysis and Machine Intelligence*, *22*(7), pp. 733–737, 2000.
22. S. Lefevre, J. Gerard, A. Piron, N. Vincent, An extended snake model for real-time multiple object tracking, *Proc. ACIVS*, pp. 268–275, 2002.
23. T. Barbu, S. -I. Bejinariu, CNN-based Moving Vehicle Recognition using GMM-based Foreground Modeling, Level-set based Segmentation and Kalman Filter-based Tracking, *18th International Conference on INnovations in Intelligent SysTems and Applications - INISTA 2024*, Craiova, Romania, pp. 1–6, September 4–6, 2024, IEEE.
24. T. Barbu, S. -I. Bejinariu, R. Luca, Automatic Vehicle Detection and Tracking Using Cascade Classifiers and CNN-Based Multi-Scale Feature Extraction, *2024 IEEE Conference on Advanced Topics on Measurements and Simulation, ATOMS 2024*, Constanta, Romania, pp. 50–53, August 28–30, 2024. IEEE.
25. M. Zhai, X. Xiang, N. Lv, X. Kong, Optical flow and scene flow estimation: A survey. *Pattern Recognition*, 114, 107861, 2021.
26. J. T. Philip, B. Samuvel, K. Pradeesh, N. K. Nimmi, A comparative study of block matching and optical flow motion estimation algorithms, *2014 Annual International Conference on Emerging Research Areas: Magnetics, Machines and Drives (AICERA/iCMMD)*, pp. 1–6, 2014. IEEE.
27. B. D. Lucas, T. Kanade, An iterative image registration technique with an application to stereo vision, *Proceedings of Imaging Understanding Workshop*, pp. 121–130, 1981.
28. A. Dosovitskiy et al., Flownet: Learning optical flow with convolutional networks, *Proceedings of the IEEE international conference on computer vision*, 2015.
29. J. Hur, S. Roth, Optical flow estimation in the deep learning age, *Modelling human motion: from human perception to robot design*, pp. 119–140, 2020.
30. N. Kumar, O. P. Verma, A. S. Parihar, Evolution of transformer-based optical flow estimation techniques: a survey. *Multimedia Tools and Applications*, pp. 1–42, 2024.
31. B. K. P. Horn, B.G. Schunck, Determining optical flow, *Artificial Intelligence*, vol 17, pp. 185–203, 1981.
32. D. Fortun, P. Bouthemy, C. Kervrann, Optical flow modeling and computation: A survey, *Computer Vision and Image Understanding*, *134*, pp. 1–21, 2015.
33. S. Aslani, H. Mahdavi-Nasab, Optical flow based moving object detection and tracking for traffic surveillance, *International Journal of Electrical, Computer, Energetic, Electronic and Communication Engineering*, *7*(9), pp. 1252–1256, 2013.
34. P. Soille, Morphological operators, *Computer Vision and Applications*, pp. 483–515, 2000, Academic Press.

35. F. Charif, Z.-e. Baarir, A Fast Modified Horn & Schunck Method, *2006 IEEE International Symposium on Industrial Electronics*, Montreal, QC, Canada, pp. 526–531, 2006.
36. S. Liu, *Object trajectory estimation using optical flow*, Utah State University, 2009.
37. Y. Shirai, Y. Mae, S. Y. Yamamoto, Object tracking by using optical flows and edges, *Robotics Research: The Seventh International Symposium*, pp. 440–447, Springer London, 1996.
38. J. Shi, Good features to track, *Proceedings of IEEE conference on computer vision and pattern recognition*, pp. 593–600, 1994. IEEE.
39. H. Salmane, Y. Ruichek, L. Khoudour, Object tracking using Harris corner points based optical flow propagation and Kalman filter, *14th International IEEE Conference on Intelligent Transportation Systems (ITSC)*, Washington, DC, USA, pp. 67–73, 2011.
40. C. Tomasi, T. Kanade, Detection and tracking of point, *Int J Comput Vis*, *9*, pp. 137–154, 3, 1991.
41. T. Barbu, S.-I. Bejinariu, R. Luca, Development of an Annotated Database for Assessing the Performance of Deep Learning-based Vehicle Detection and Tracking Models, *Annals of the Academy of Romanian Scientists, Series on Science and Technology and Information*, Publishing House of the Academy of Romanian Scientists, Vol. 17, No. 2, pp. 22–30, 2024.
42. S. Jia, Y. Cao, F. Yang, X. Lu, X. Lu, Multi-object Tracking by Detection and Query: an efficient end-to-end manner, *arXiv preprint* arXiv:2411.06197, 2024.
43. T. Barbu, Deep Learning-based Multiple Moving Vehicle Detection and Tracking using a Nonlinear Fourth-order Reaction-Diffusion based Multi-scale Video Object Analysis, *Discrete & Continuous Dynamical Systems - Series S, AIMS Journals*, 16 (1), pp. 6–32, 2023.
44. T. Barbu, Multiple Pedestrian Tracking Framework using Deep Learning-based Multiscale Image Analysis for Stationary-camera Video Surveillance, *2022 IEEE International Smart Cities Conference (ISC2)*, pp. 1–7, Sept. 2022, IEEE.
45. J. L. Lions, *Quelques methods de resolution des problèmes aux limites nonlinéares,* Dunod. Gauthier-Villars, Paris, 1969.
46. P. Johnson, *Finite Difference for PDEs*, School of Mathematics, University of Manchester, Semester I, 2008.
47. L. Wen, D. Du, Z. Cai, Z. Lei, M. Chang, K. Qi, ... & S. Lyu, UA-DETRAC: A new benchmark and protocol for multi-object detection and tracking, *Computer Vision and Image Understanding*, *193*, 102907, 2020.
48. L. Wang, Y. Lu, H. Wang, Y. Zheng, H. Ye, X. Xue, Evolving boxes for fast vehicle detection, *2017 IEEE International Conference on Multimedia and Expo (ICME)*, pp. 1135–1140, 2017.
49. P. Dollár et al., Pedestrian detection: A benchmark, *2009 IEEE conference on computer vision and pattern recognition*, 2009. IEEE.
50. C. Szegedy, V. Vanoucke, S. Ioffe, J. Shlens, Z. Wojna, Rethinking the inception architecture for computer vision, *Proc. of IEEE Conference on Computer Vision and Pattern Recognition*, pp. 2818–2826, 2016.
51. Z. Wu, C. Shen, A. Van Den Hengel, Wider or deeper: Revisiting the resnet model for visual recognition, *Pattern Recognition*, 90, pp. 119–133, 2019.
52. A. Dominguez-Sanchez, M. Cazorla, S. Orts-Escolano, Pedestrian movement direction recognition using convolutional neural networks, *IEEE Trans. on Intel. Transport. Systems*, 18(12), pp. 3540–3548, 2017.
53. JAAD Dataset - https://data.nvision2.eecs.yorku.ca/JAAD_dataset/
54. M. A. B. Zuraimi, F. H. K. Zaman, Vehicle detection and tracking using YOLO and DeepSORT, *11th IEEE Symposium on Computer Applications & Industrial Electronics (ISCAIE)*, pp. 23–29, April 2021.

Chapter 7
Conclusions

The applications of the mathematical models based on nonlinear partial differential equations in the image processing and analysis, computer vision and artificial intelligence domains have been addressed in this book. The work has surveyed the existing research in these fields, but mainly insisted on the author's most important contributions in them.

As already mentioned, the research results described here represent a continuation of those presented in our previous Springer book, which was published in 2019 and treated only the PDE-based image denoising and restoration and inpainting domains. Besides the results achieved by us in these two fields since 2019, this book also describes our research in some important image analysis and computer vision domains, such as image compression, static and video image segmentation, content-based image recognition, indexing and retrieval, and video object detection and tracking.

Only the image processing and analysis sub-domains approached in our past works have been considered here. Since we have conducted a high volume of research in these PDE-based image processing and analysis areas in the last two decades, the scientific results discussed in this book have been disseminated in many research articles published in recognized international journals and international conference volumes, and book chapters. These published works are listed in the reference sections of the chapters and cited widely throughout the manuscript. However, the cited papers have not been provided equal treatment in this book. Thus, some large presentations have been provided for those disseminating the most important theoretical and practical results, while others have been discussed more briefly. Also, in some cases only those parts of the paper that involve PDE models are addressed here, and not the entire work.

Each chapter of this book, except the introduction and conclusion chapters, was concerned with one of the mentioned research domains. After surveying the main nonlinear PDE-based approaches in that field, our own contributions have been presented. They represent effective image processing or analysis techniques

T. Barbu, *Digital Image Processing, Analysis and Computer Vision Using Nonlinear Partial Differential Equations*, Studies in Computational Intelligence 1211,
https://doi.org/10.1007/978-3-031-89576-0_7

based on novel variational or nonlinear anisotropic and reaction–diffusion models, whose results have been described and compared to those of the well-known existing approaches.

Rigorous mathematical treatments have been provided for some of these nonlinear second and fourth-order PDE models, their validity (well-posednesss) being investigated. However, not all the mathematical investigations performed by us for the PDE-based models described here have been presented in this book. In some cases the reader has been referred to the papers detailing those mathematical treatments.

Some fast-converging approximation algorithms that solve numerically the proposed PDE models and are stable and consistent to them have been developed by applying the finite difference method. Not all of them have been provided in detail here, since some of them are quite similar to other already described discretization schemes. Also, some variational approaches that generate second or fourth-order nonlinear diffusion-based models by minimizing various energy functionals have not been detailed here, the reader being referred to the published works describing them.

While we have our own contributions in all the research domains approached in this work, some chapters describe many more PDE-based techniques developed by us than others. Thus, we have brought more contributions to the image and video denoising and restoration, inpainting and multi-scale image analysis domains, than to other research fields.

The research results described in this book show once more the usefulness of the nonlinear partial differential equations in all the image processing and analysis and computer vision fields, and how they can be combined successfully to the CV and AI models based on machine and deep learning algorithms. The nonlinear diffusion-based approaches proposed by us have been applied effectively in combination to those computer vision solutions.

So, the nonlinear PDE-based filtering techniques proposed in the second chapter for 2D images, video images and multi-layer images, and dealing with various types of noise, represent effective pre-processing solutions that improve the analyzed image quality and facilitate the further computer vision tasks. The reconstruction techniques using the proposed PDE-based inpainting models, described in the third chapter, also improve the images by removing the missing regions and undesired entities from them, thus facilitating the next image analysis processes. Also, the images enhanced by these pre-processing methods have been used successfully by the machine and deep learning-based models whose training and validation procedures require pre-processed image datasets.

The second- and fourth-order nonlinear diffusion-based multi-scale analysis algorithms that are described in the fourth, fifth and the sixth chapters of this book are have provided good results for various image analysis and computer vision tasks when combined to machine learning schemes and deep neural networks. Thus, some PDE-based scale-space representations have been successfully used in combination to some deep learning-based high-level feature extraction schemes for solving the content-based image indexing and retrieval, temporal video segmentation and video object tracking-by-detection tasks presented here. Other diffusion-based multi-scale spaces created by us, which have been combined to various feature extractors applied at multiple scales, have been used successfully for texture and content-based image recognition and also in the video object tracking frameworks that use the detections generated by various machine learning algorithms and convolutional neural networks.

The introduced nonlinear PDE-based segmentation techniques, obtained from edge-enhancing diffusion-based filtering and multi-scale analysis, have also provided some important results for the computer vision tasks. Thus, the anisotropic diffusion-based edge-based segmentation approaches described in this book and used for image compression could also be applied successfully to many other CV and AI tasks. The image segmentation algorithms using the nonlinear anisotropic diffusion-based geodesic active contour models proposed by us have been combined successfully to some machine learning-based detection techniques. So, the GAC-based segmentation approaches described in the fifth chapter improve considerably the vehicle detection results provided by the adaptive boosted cascade classifiers and GMM-based schemes.

There are some other image processing and analysis fields where the variational and PDE-based models have been successfully applied, but which have not been discussed in this book. They include domains such as the image registration or the 3D object recognition and we aim to approach them as part of our future research in this area.

The manufacturer's authorised representative in the EU is Springer Nature Customer Service Centre GmbH, Europaplatz 3, 69115 Heidelberg, Germany. If you have any concerns regarding our products, please contact ProductSafety@springernature.com

Printed and bound by CPI Group (UK) Ltd, Croydon, CR0 4YY
30/09/2026
02225471-0012